T0181405

Advances in Intelligent Systems and Computing

Volume 1313

The series "Advances in Intelligent Systems and Computing" contains publications on theory, applications, and design methods of Intelligent Systems and Intelligent Computing. Virtually all disciplines such as engineering, natural sciences, computer and information science, ICT, economics, business, e-commerce, environment, healthcare, life science are covered. The list of topics spans all the areas of modern intelligent systems and computing such as: computational intelligence, soft computing including neural networks, fuzzy systems, evolutionary computing and the fusion of these paradigms, social intelligence, ambient intelligence, computational neuroscience, artificial life, virtual worlds and society, cognitive science and systems, Perception and Vision, DNA and immune based systems, self-organizing and adaptive systems, e-Learning and teaching, human-centered and human-centric computing, recommender systems, intelligent control, robotics and mechatronics including human-machine teaming, knowledge-based paradigms, learning paradigms, machine ethics, intelligent data analysis, knowledge management, intelligent agents, intelligent decision making and support, intelligent network security, trust management, interactive entertainment, Web intelligence and multimedia.

The publications within "Advances in Intelligent Systems and Computing" are primarily proceedings of important conferences, symposia and congresses. They cover significant recent developments in the field, both of a foundational and applicable character. An important characteristic feature of the series is the short publication time and world-wide distribution. This permits a rapid and broad dissemination of research results.

Indexed by DBLP, EI Compendex, INSPEC, WTI Frankfurt eG, zbMATH, Japanese Science and Technology Agency (JST), SCImago.

All books published in the series are submitted for consideration in Web of Science.

More information about this series at http://www.springer.com/series/11156

Davor Sumpor · Kristian Jambrošić ·
Tanja Jurčević Lulić · Diana Milčić ·
Ivana Salopek Čubrić · Irena Šabarić
Editors

Proceedings of the 8th International Ergonomics Conference

ERGONOMICS 2020

Editors
Davor Sumpor
Faculty of Transport and Traffic Sciences
University of Zagreb
Zagreb, Croatia

Tanja Jurčević Lulić
Faculty of Mechanical Engineering
and Naval Architecture
University of Zagreb
Zagreb, Croatia

Ivana Salopek Čubrić
Faculty of Textile Technology
University of Zagreb
Zagreb, Croatia

Kristian Jambrošić
Faculty of Electrical Engineering
and Computing
University of Zagreb
Zagreb, Croatia

Diana Milčić
Faculty of Graphic Arts
University of Zagreb
Zagreb, Croatia

Irena Šabarić
Faculty of Textile Technology
University of Zagreb
Zagreb, Croatia

ISSN 2194-5357 ISSN 2194-5365 (electronic)
Advances in Intelligent Systems and Computing
ISBN 978-3-030-66939-3 ISBN 978-3-030-66937-9 (eBook)
https://doi.org/10.1007/978-3-030-66937-9

This Springer imprint is published by the registered company Springer Nature Switzerland AG
The registered company address is: Gewerbestrasse 11, 6330 Cham, Switzerland

Committees

The Conference is **organized by**:

The Conference is **endorsed by**:

Conference Endorsed by the
IEA
International Ergonomics Association

Hrvatsko akustièko društvo

HAD

Acoustical Society of Croatia

The Conference is **co-organized by**:

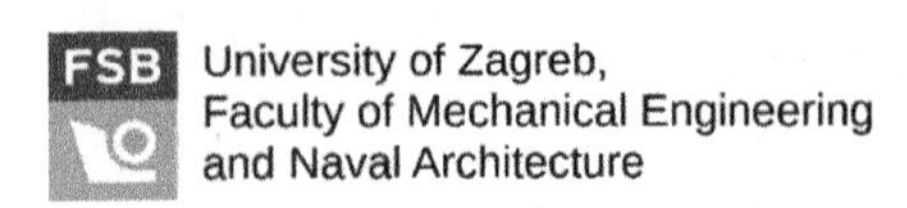

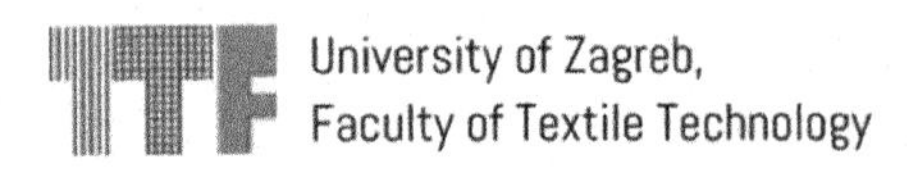

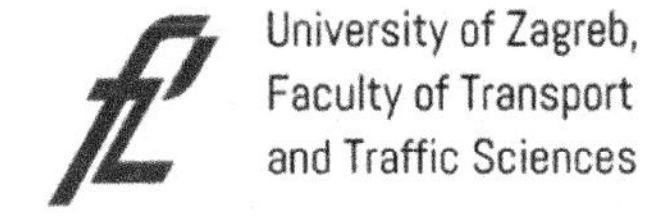

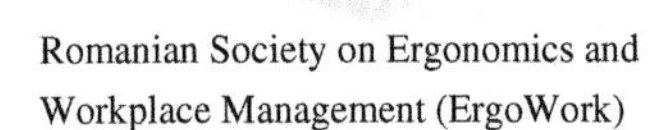

International Scientific Committee

Organizing Committee

Davor Sumpor, Croatia (Organizing Committee Chair, General Chair)
Beata Mrugalska, Poland
Szabó Gyula, Hungary
Tino Bucak, Croatia
Goran Čubrić, Croatia
Anca Draghici, Romania
Tanja Jurčević Lulić, Croatia
Marjan Leber, Slovenia
Jasna Leder Horina, Croatia
Elma Mulaomerović, Taiwan
Ivana Salopek Čubrić, Croatia
Irena Šabarić, Croatia
Emilija Zdraveva, Croatia

Programme Committee

Tanja Jurčević Lulić, Croatia (Programme Committee Chair)
Kristian Jambrošić, Croatia
Diana Milčić, Croatia
Ivana Salopek Čubrić, Croatia
Irena Šabarić, Croatia
Eric Min-Yang Wang, Taiwan

Technical Committee

Jasna Leder Horina, Croatia (Technical Committee Chair)
Eva Bolčević, Croatia
Daria Ćurko, Croatia
Božena Jurčić, Croatia
Emilija Zdraveva, Croatia

Preface

The *8th International Ergonomics Conference—ERGONOMICS 2020* was held from 2 to 5 December 2020 at the University Campus Borongaj "ZUK Borongaj", in Zagreb, the capital of Croatia.

Ergonomics in Croatia has existed as a scientific and professional discipline for several decades. Let us remember that it all started in 1974 when Prof. Emeritus Dragutin Taboršak, Ph.D., the founder and the first president of the Croatian Ergonomics Society (CrES), established CrES in Croatia. The achievements as well as recent and relevant ideas in the field of ergonomics have been continuously discussed and exchanged with the international scientific community at the Conferences in the series "Ergonomics", along with other means of communication and networking.

The Conferences in the series "Ergonomics" have been organized by the (CES) from 2001 as part of CES objectives to promote ergonomics and exchange knowledge and experience with the scientific and professional community from Croatia and the world. The Conference traditionally brings together enthusiasts, experts and scientists from Croatia and from all over the world, and they bring their know-how to the table.

This time, the Conference *ERGONOMICS 2020* is a joint project with our co-organizing partners:

- CAES (Chinese Association of Ergonomics Societies)
- CES (Chinese Ergonomics Society)
- ErgoWork (Romanian Society on Ergonomics and Workplace Management)
- EST (Ergonomics Society of Taiwan)
- HKES (Hong Kong Ergonomics Society)
- MET (Hungarian Ergonomics Society)
- FPZ (Faculty of Transport and Traffic Sciences), University of Zagreb, Croatia
- FSB (Faculty of Mechanical Engineering and Naval Architecture), University of Zagreb, Croatia
- TTF (Faculty of Textile Technology), University of Zagreb, Croatia

Furthermore, the Conference has been this time endorsed by:

- IEA (International Ergonomics Association)
- FEES (Federation of European Ergonomics Societies)
- ASC (Acoustical Society of Croatia)

The Conference programme this time included more types of oral presentations of papers from the following groups of topics (not limited):

- Aesthetics and Ergonomics
- Biomechanics and Modelling in Ergonomics
- Cognitive Ergonomics
- Education and Trainings in Work Safety and Ergonomics
- Ergonomics for People with Disabilities and Aging Population
- Ergonomics in Product and Process Desig;
- Ergonomic Regulations, Standards and Guidelines
- Healthcare Ergonomics
- Physical Ergonomics and Human Factors
- Human Comfort
- Safety and Risk Ergonomics
- Psychoacoustic Ergonomics
- Social and Occupational Ergonomics
- Traffic and Transport Ergonomics

The Organizing Committee (OC) of the *ERGONOMICS 2020 Conference* received more than 45 contributions within a diverse range of conference topics. All submissions have been peer-reviewed by the International Scientific Committee and referees from abroad and Croatia, regardless of what type of oral presentation has been chosen (live oral presentation on the spot, real-time online presentation, pre-recorded lecture).

The participants have come from the following countries: Austria, Croatia, Brazil, Bulgaria, Bosnia and Herzegovina, Greece, Hungary, India, Italy, Poland, Portugal, Romania, Russian Federation, Ukraine, Slovenia, Spain and Taiwan ROC.

During the Conference opening ceremony, Davor Sumpor (President of CrES and Conference Chair) invited Tomislav Mlinarić (FPZ Dean) and Jose Orlando Gomes (IEA Vice-President) to address the participants with a few words of welcome.

In the introductory part of the Conference sessions, the following invited speakers contributed with their lectures: Jose Orlando Gomes (Brazil), Eric Min-Yang Wang (Taiwan ROC), Abhijit Majumdar (India), Nejc Šarabon (Slovenia), Beata Mrugalska (Poland) and Ivana Salopek Čubrić (Croatia).

Due to the global and local situation caused by the COVID-19 virus pandemic, in the circumstances of a situation called "New normal", the Conference was held as a mixed Conference which included the following types of oral presentations: live oral presentation on the spot, real-time online presentation, pre-recorded lecture (only voice).

The Conference provided a great opportunity for all the participants and all stakeholders from Croatia and abroad to contribute to the advances in ergonomics in Croatia once again all the way through to valuable exchange of knowledge and experience.

I would like to express deep appreciation to all co-organizing partners, patrons and sponsors, reviewers and all the members of all the Conference bodies, authors of the papers and Conference participants, who have all together enabled and helped to make this Conference in the "Ergonomics" series successful once again.

Finally, I am pleased to note that even in the circumstances of the pandemic situation called "New normal", CrES has been receiving continuous active support of our constant and reliable international partners, who we are very proud of, and for which we are very grateful.

Zagreb, Croatia

Assoc. Prof. Davor Sumpor, Ph.D.
President of the Croatian Ergonomics Society
Conference Chair

The Conference provided a great opportunity for [illegible] participants and all stakeholders from Croatia and abroad to contribute to the [illegible] in Croatia once again and [illegible] toward a [illegible] exchange of knowledge and experience.

I would like to express [illegible] to all the organizers [illegible] and sponsors, reviewers and [illegible] of the papers [illegible] Conference [illegible] [illegible] [illegible]

[illegible]

Acknowledgements

The 7th International Ergonomics Conference *ERGONOMICS 2018—Emphasis on Wellbeing* took place from 13 to 16 June 2018 in Zadar, Croatia, and it was held again in a relaxed atmosphere of the Falkensteiner Hotels & Residences, Borik. It was our great pleasure to see some of our colleagues once again in Zadar.

The Conferences in this series have been organized by the **Croatian Ergonomics Society (CrES)** every three years since 2001, and from 2016 every two years, as part of CrES objectives to promote ergonomics and exchange knowledge and experience with the scientific and professional community from Croatia and the world, in their consecutive order, as follows:

- 1st International Ergonomics Conference—*Ergonomics 2001*, Zagreb
- 2nd International Ergonomics Conference—*Ergonomics 2004*, Stubičke Toplice, Zagreb
- 3rd International Ergonomics Conference—*Ergonomics 2007*, Stubičke Toplice, Zagreb
- 4th International Ergonomics Conference—*Ergonomics 2010*, Stubičke Toplice, Zagreb
- 5th International Ergonomics Conference—*Ergonomics 2013*, Zadar
- 6th International Ergonomics Conference—*Ergonomics 2016*—Focus on Synergy, Zadar.

At that time, *ERGONOMICS 2018—Emphasis on Wellbeing* was a joint project with:

- **Ergonomics Society of Taiwan** (EST)
- **Chinese Ergonomics Society** (CES)
- **Hong Kong Ergonomics Society** (HKES),
- **Russian Inter-Regional Ergonomics Association** (IREA)
- **Faculty of Transport and Traffic Sciences** (FPZ), University of Zagreb

- **Faculty of Mechanical Engineering and Naval Architecture** (FSB), University of Zagreb and
- **Faculty of Textile Technology** (TTF), University of Zagreb.

Furthermore, *ERGONOMICS 2018—Emphasis on Wellbeing* was then endorsed by:

- **International Ergonomics Association** (IEA)
- **Federation of the European Ergonomics Societies** (FEES)
- **Acoustical Society of Croatia** (ASC).

During the Conference opening ceremony, Davor Sumpor (President of CrES and Conference Chair) emphasized the importance of the Conference as a place of knowledge exchange and experience exchange between scientists and experts from Croatia and the World and invited Jose Orlando Gomes (IEA Vice-President) and Szabo Gyula (Chair of FEES Communication and Promotion Committee) to address the participants with a few words of welcome.

More than 70 papers had been submitted, and after the review process, 50 papers were presented at the Conference in oral or poster sessions. The presented papers have been published in the Conference Book of Proceedings (USB), which was printed with full texts of all the reviewed papers only. The participants came from the following countries: Austria, Canada, Croatia, Brazil, Germany, Greece, Hungary, India, Indonesia, Iran, Nigeria, Mexico, Poland, Portugal, Romania, Republic of Korea, Russian Federation, Slovenia, South Africa, Switzerland, Taiwan ROC and the USA.

In the introductory part of the Conference sessions, the following invited speakers contributed with their lectures: Denis Coelho (Portugal), Maggie Graf (Switzerland), Eric Min-Yang Wang (Taiwan) and Uwe Reischl (USA).

For the first time the participants from several countries in the region (Croatia, Slovenia, Hungary, Austria, Russia and Poland) were able to attend the day-long CREE—IEA—FEES workshop "An Approach to Growing Ergonomics in Europe", which can be equated to 1 ECT of training in Professional Affairs for CREE certification. Six participants from Croatia attended the workshop. The feedback from all participants was quite useful and encouraging.

Conference events such as Welcome party, Zadar City Tour, Surprise excursion to the lake Vrana in Pakoštane and the Gala Party were additional opportunities to open up new possibilities, meet old and make new friends.

For all the participants and stakeholders, the Conference was a chance to contribute to a valuable exchange of knowledge and experiences, to encourage stronger interconnections between science and business, as well as to contribute to the advances in ergonomics.

I am pleased to note that the successful Conference was an outcome of the work of a large group of people. We would like to express deep appreciation to all co-organizing partners, patrons and sponsors, who enabled and helped to make this Conference successful. We also acknowledge the authors themselves, who contributed to the Conference success with their expertise, appreciating authors′

willingness to devote their time for writing the papers and conducting the studies. Finally, with special respect, I would like to thank sincerely all the reviewers and all the members of all the Conference bodies from Croatia and abroad for their devoted work usually in their leisure time which was in the last part of their tasks under huge time pressure, as well as their families or partners for their patience during the preparation of the Conference in the several past months.

I believe that, with the help of international partners, CrES will be able to keep up this level of performance in the future because of the involvement of several young colleagues from Croatia in the Conference organization, for which I am very happy and grateful.

We are looking forward to meeting our dear friends and colleagues in Croatia in 2020 once again, and you are very welcome to inform the others who might be interested to participate in the *8th International Ergonomics Conference—ERGONOMICS 2020.*

On behalf of the Croatian Ergonomics Society,

Organizing Committee of the *Ergonomics 2018—Emphasis on Wellbeing*

willingness to devote [illegible] the papers and [illegible] the studies. Finally, with special [illegible] to thank sincerely all the reviewers and all the members of all the [illegible] bodies from Greece and abroad [illegible] in their [illegible] the last part of their task under the time pressure, as well as the [illegible] preparation of the Conference [illegible]

I believe that, with the [illegible] level expert [illegible] young colleagues from [illegible]

[illegible]

In Memoriam

Professor Emeritus Dragutin Taboršak (12 June 1925–20 December 2018)
It is with deep sorrow that we have to inform that our dear colleague, Professor Emeritus Dragutin Taboršak, member of the Department of Industrial Engineering at the Faculty of Mechanical Engineering and Naval Architecture, University of Zagreb, as well as an honorary member of the Croatian Ergonomics Society, passed away peacefully. He has left behind many close friends and colleagues. His death touched us deeply and made us very sad.

Professor Emeritus Dragutin Taboršak was mentor, teacher and colleague who gave his enthusiasm, personality and freshness to his work. He inspired his students by giving them selflessly his knowledge, experience and his time working with them both in the classroom and in the laboratory.

Dragutin Taboršak was born in Sunja and in his early teenage years went to Zagreb for high school education, which he completed in 1944. His beginnings in the technical profession was working in his father's locksmith workshop and parallel he started studying at the Technical Faculty at the University of Zagreb. He completed his study in 1950 and afterwards started to work in industry, first in the company "Rade Končar" and afterwards in the company ZET and later in the company "Radnik". In all these companies, he was dealing with organization of preparation of work as well as production management. The experience obtained in these companies showed him how much can still be investigated and implemented in this field.

Therefore, he went back to the Faculty to do his doctoral thesis and more research in the field of organizing the preparation of work. He finished his Ph.D. and became assistant professor at the Higher Post-secondary Technical School at the University of Zagreb in 1962, and soon after, he became associate professor. From 1966 till 1968, he was the Dean. In 1968, the Higher Technical School became part of the Faculty of Mechanical Engineering and Naval Architecture where Professor Taboršak was Dean from 1968 to 1970. During his academic career, he was part of many Faculty and University activities as well as councils. It should be highlighted that he was head of doctoral council where he developed new fields of research and introduced ergonomics as a new scientific field in Croatia.

During his Ph.D. study, Prof. Taboršak spent 8 months in Paris in companies "Bureau des Temps Elementaires" and "Renault" as well as 3 months in Darmstadt in the company "REFA". In these periods, he had the opportunity to learn about work ergonomics and bring that knowledge to Croatia, implementing it in his lectures.

Collecting all this knowledge and experience from abroad and in industry he came to the conclusion that ergonomics needed to be more implemented and used in Croatia; therefore, he established the Croatian Society of Ergonomics on 20 May 1974 in Zagreb where he was the first president. Later, the society changed its name to the Croatian Ergonomics Society (CrES). In accordance with the mission and vision of the founder of the society, CrES has taken care of scientific, professional and educational fields of ergonomics in the Republic of Croatia, as well as promoted cooperation with the business community. Since the establishment, CrES has been interconnected with relevant societies and organizations in the field of ergonomics worldwide to sustain the scientific and professional level.

The main scientific interests of Prof. Taboršak were Work Studies, Ergonomics and Production Organisation about which he published many scientific and professional papers and also did many seminars in industry. His knowledge in this field he implemented in teaching and established the Laboratory for work studies, first of that kind in Croatia. Also, he published the book Work Studies that had four editions and this book made the foundations in the field of ergonomics in Croatia, and it is still in use.

As a respectable professor responsible for developing new research areas, for his contribution in science and teaching, mentoring students, for participation in many scientific and industrial project solving real production problems Prof. Dragutin Taboršak received in 1992 from the Republic of Croatia the Life Achievement Award.

There is so much that can be told about the rich history of the activities of Professor Taboršak, but we are going to remember Prof. Taboršak most for his optimistic character, creativity in his work and devotion to his Faculty and country of Croatia. With the death of Professor Emeritus Dragutin Taboršak, the professional and scientific community lost a prominent member, extraordinary expert in the field of work study and ergonomics, a dear colleague and friend who will be remembered by generations of students, colleagues and friends who worked with him.

May he rest in peace!

Contents

Invited Lectures

Approach Towards Design of Functional Sportswear for Improved Human Performance

Ivana Salopek Čubrić

Abstract Contemporary design and production of textile and clothing is focused on development of products with innovative characteristics and hi-tech functionalization for improved human performance. Therefore, an important innovation area is the R&D of specifically engineered materials and assemblies. The research is directed towards delivery of a pre-defined performance or functionality to the user, over and above its normal functions. This paper gives review of the state-of-the-art research in the area of functional sportswear design and discusses niches for the innovations beyond. The experimental part covers evaluations within three crucial segments, i.e. thermophysiological comfort of material observed trough the permeability index, thermophysiological comfort of single layered sportswear observed using thermal camera and sensorial comfort of single layered sportswear material. The test results are supported by regression models. Throughout the discussion and conclusion is outlined the need for comprehensive and systematic approach towards design and further evaluation of sportswear in order to enhance and augment body functions.

Keywords Functional · Sportswear · Temperature · Permeability index · IR thermography

1 Introduction

Contemporary textile and clothing producers are focused in development of products with innovative characteristics and hi-tech functionalization that can represent an important added value. The properties and the characteristics that were initially developed for products for special use are nowadays often present in textiles for everyday use which can be distinguished by multiple functions.

I. Salopek Čubrić (✉)
Department of Textile Design and Management, University of Zagreb Faculty of Textile Technology, Zagreb, Croatia
e-mail: ivana.salopek@ttf.unizg.hr

D. Sumpor et al. (eds.), *Proceedings of the 8th International Ergonomics Conference*, Advances in Intelligent Systems and Computing 1313,
https://doi.org/10.1007/978-3-030-66937-9_1

Therefore, new textile products for improved human performance are an important innovation area in the research and production of textile and clothing like sportswear and protective wear.

The term functional clothing refers to different types of clothing or assemblies, specifically engineered to deliver a pre-defined performance to the final user. Functional clothing may be classified according to the main application areas:

I. Sports clothing
The clothing is engineered with regards to the specifics of each sport, for example golf, tennis, soccer, football, basketball, baseball, swimming, diving, running, skiing, ice skating, cycling, motorcycle riding, fencing, martial arts, fitness, alpine climbing, etc.
II. Protective clothing
The clothing gives protection against mechanical impact, physical injury, drowning, heat and fire, cold, rain, electric shock, radiation, dangerous substances, etc.
III. Medical-functional clothing
This category includes healthcare/hygiene clothing, surgical clothing, therapeutic clothing and intelligent clothing.
IV. Clothing for special needs
The category includes clothing which main function is improving the quality of life for people with disabilities or special needs, like paraplegics, handicapped, the elderly, Alzheimer's sufferers, arthritis patients, incontinence sufferers, etc.

2 Aspects of Sportswear Comfort

In order to provide a comfortable microclimate for a specific end-user, designers must define specifics for each type of functional clothing as well as to obtain feedback from the practitioner regarding specific issues. Table 1 gives overview of the demands of body and sport that need to be calculated when designing functional sportswear.

According to the recent reports [1], the global sportswear market is valued at US $ 84.1 Bn in the year 2018 and is expected to touch a valuation of US$ 108.7 Bn by end of year 2025. This gives a clear picture of how important it is to direct scientific interest towards the sportswear improvement. The following segments have to be combined within this mission:

(i) optimal material characteristics,
(ii) optimized heat and moisture management through material and
(iii) increased comfort of garment.

Comfort of sportswear is a complex phenomenon that should be observed from three different main aspects:

Table 1 Demands for the design of functional sportswear

Demands of body				
Thermophysiological regulation	Protection	Anthropometry	Movement	Psychological considerations
Heat and mass transfer, insulation, environmental conditions, workload	Contact/ non-contact sport, environmental conditions, health, safety	Body measures	Postures, agility	Style, appearance, perception
Demands of sport				
Safety		Conditions		
Protection, identification, rules		Duration of activity, location, season, climate		

I. Thermophysiological comfort
Thermophysiological comfort comprises transport processes through the sportswear. Key notions include thermal insulation, heat resistance, moisture management and breathability.

II. Sensorial comfort
Sensorial comfort characterizes the mechanical sensations that could be evaluated using objective or subjective methods.

III. Psychological comfort
The psychological comfort is rather subjective. It is affected by personal preferences, fashion, different ideology, etc.

2.1 Thermophysiological Comfort of Material

For the decades, the major innovation in the single layer sportswear was related to the introduction of new fibers. Fabrics for sportswear also need to be specifically constructed in terms of geometry of yarn and construction of fabric, all in order to ensure enhanced performance of the wearer [2]. The above indicates the importance of further investigation of mechanical functionalization and the need to apply new approaches to engineer innovative structures [3]. It is a joint point of view that further efforts of researchers need to be given towards investigation of optimal spinning technique and sophisticated fibers to directly affect the functionality of fabrics used for targeted sportswear. A number of investigations have been carried out to describe the effect of fabric structural parameters on different fabric properties, such as thermal resistance, vapour resistance, air permeability and mechanical properties [4–9]. However, the relative contributions of each of these segments to the comfort and performance of sportswear and sporting activities remain insufficiently defined. It is a joint point of view that further studies should be conducted in order to valorize the influence of yarn and fabric parameters on functionality of materials for specific sportswear. The transfer of heat and moisture trough sportswear materials is one of the most important properties affecting human perception of comfort. In actual wearing circumstances, it is impossible to separate heat and moisture transfer through sportswear what brings out the necessity of assessing the combination of the evaporation and heat transfer.

2.2 Thermophysiological Comfort of Clothing

The knowledge and investigation of human physiology creates a crucial basis for the development of sportswear with added value that facilitates optimal performance of athletes. Over the last decade, much attention has been paid to regional thermoregulatory effectors responses in order to adapt the choice and location of material to better meet the body needs. It is confirmed that infrared thermography can be used to assess the efficiency of the clothing intervention and may be promising in both evaluating normal patterns and deviations from these patterns with body mapping whole-body or specific areas [10–14]. Preliminary results of body mapping garment designed for the upper body were promising for improved thermal comfort. The individual qualitative and quantitative information obtained by infrared thermography can be extremely valuable to assess the efficiency of specific garments at the skin interface (warming, cooling, reducing temperature contrasts), consequently leading to improvements in thermal comfort and human performance. With careful data collection and standardized undressing procedures, this opens a wide area of testing the influence of many kinds of different sportswear garments. Further studies are required to understand the influence of body mapping sportswear in the cold and development of more customized garments. Also, a good compromise must be found between heat and vapour resistance. Investigation confirmed that manipulation of local heat resistances could be combined with the manipulation of vapour resistances to maximize comfort and performance in moderate to high intensity activities where sweat production and skin wettedness are extremely important. This should lead to body mapping garments customized by function, what is additional value for the development of personalized sportswear. The knowledge of the thermal patterns under sport apparels could be extended to a myriad of clothing conditions and exercise types in order to map the similarities and differences in skin temperature distribution.

2.3 Sensorial Comfort

Consumer purchases of textiles and clothing are very often driven by visual appearance and sensory attraction, especially consumers' visual sense and sense of touch [14]. Touching a fabric is among first actions that consumers perform in order to evaluate fabric properties in order to choose a suitable fabric for garments and to estimate the performance of the fabric for its end use. Therefore, the evaluation of materials using sensory analysis has drawn much international attention [15–17]. Regarding the subjective perception of sensorial and thermal comfort, the outcomes indicated importance of human feedback in order to develop materials and clothing for enhanced performance, as well as important discrepancies in sensorial perceptions between volunteers with different demographic profile.

3 Experimental

The Experimental part of this paper covers three aspects important for the design of functional single layered sportswear, i.e.: thermophysiological comfort of material observed trough the permeability index, thermophysiological comfort of single layered sportswear observed using thermal camera and sensorial comfort of single layered sportswear material.

The permeability index, as a measure of the efficiency of evaporative and heat transport in a clothing system is a good indicator of transfer properties for porous, single layered fabrics. The permeability index (i_m) is defined as the relationship between thermal resistance and evaporative (water vapour) resistance, what is defined by the following equation:

$$i_m = 0.060 \cdot \frac{R_{ct}}{R_{et}} = 0.060 \cdot \frac{(T_s - T_a) \cdot \frac{A}{H_c} - R_{ct0}}{(p_s - p_a) \cdot \frac{A}{H_e} - R_{et0}} \quad (1)$$

where i_m—permeability index; R_{ct}—fabric heat resistance, R_{et}—fabric water vapour resistance, T_s—temperature at the plate surface, T_a—the air temperature, A—area of the plate test section, H_c—the power input during the R_{ct} measurements, R_{ct0}—bare plate heat resistance, P_s—water vapor pressure at the plate surface, P_a—water vapor pressure in the air, H_e—the power input during the R_{et} measurements, R_{et0}—bare plate water vapour resistance.

In the Experimental part, permeability index is determined using the values of water vapour and heat resistance of knitted fabrics worn as single layer that are measured on the sweating guarded hotplate [5, 6]. The value of permeability index is further correlated to fabric main properties, i.e. surface mass, thickness area and porosity. The results are presented in the form of linear regression models for all observed fabrics and separately for fabrics produced of natural and regenerated fibers.

In order to improve thermophysiological comfort of single layered sportswear, the temperature of athlete's upper body is measured after training session. The measurement is conducted using IR thermal camera with thermal sensitivity <0.06 °C and accuracy of reading ±2%. The data are processed in professional software for thermal analysis and given as average for each of 22 defined body zones.

For the evaluation of sensorial comfort of single layered materials, a group of 140 evaluators is recruited. Evaluators are asked to rank seven presented fabrics for the attributes of softness and smoothness on a scale 1–10 [15].

4 Results and Discussion

The results of modeling the relationship between a dependent variable (permeability index) and explanatory variables (surface mass, thickness and porosity) using linear regression are presented in Table 2. The changes of temperature on the surface of single layered sportswear due to activity are given in Fig. 1. In Fig. 2 are presented average ranks for smoothness and softness for seven evaluated fabrics. As can be seen from Table 2, permeability index is in the range 0.29–0.40. The results of linear regression indicate that R^2 value is high for explanatory variables surface mass and thickness in the case when the natural and regenerated materials are observed separately. Regarding the variable of porosity, R^2 value 0.9979 is obtained for group of regenerated materials only.

The results presented in Fig. 1 point out important body zones with highest change of body temperature after intensive sports activity. According to the results, this refers to the zones of abdomen, lower back and part of upper back. The results of thermographic measurements should be perceived as valuable for the definition of garment construction and should be conducted for each sport separately.

As can be seen from the Fig. 2, fabrics produced of natural fibers are perceived less soft and less smooth in comparison with other investigated fabrics. The fabric mass is perceived as negative factor for the evaluation of both smoothness and softness. In the case of smoothness, the fabric produced of regenerated fibers is highly ranked, and in the case of softness, fabrics with addition of elastane. All discussed segments of evaluation should simultaneously be used when designing sportswear in order to enhance performance and body functions.

Table 2 The results of modeling

Material	Permeability index	Exp. variable	Model	R-squared
Nat.	0.29–0.37	Surface mass	y = 0.0015x + 0.1494	0.9741
Reg.	0.35–0.40	Surface mass	y = 0.0011x + 0.2377	0.9979
Nat. + Reg.	0.29–0.40	Surface mass	y = 0.0014x + 0.0174	0.6945
Nat.	0.29–0.37	Thickness	y = 0.6110x + 0.1116	0.9766
Reg.	0.35–0.40	Thickness	y = 0.6429x + 0.1443	0.9812
Nat. + Reg.	0.29–0.40	Thickness	y = 0.6032x + 0.1364	0.5601
Nat.	0.29–0.37	Porosity	y = 1.3333x + 1.3700	0.2667
Reg.	0.35–0.40	Porosity	y = 0.0011x + 0.2377	0.9979
Nat. + Reg.	0.29–0.40	Porosity	y = 0.0004x + 0.3279	0.4869

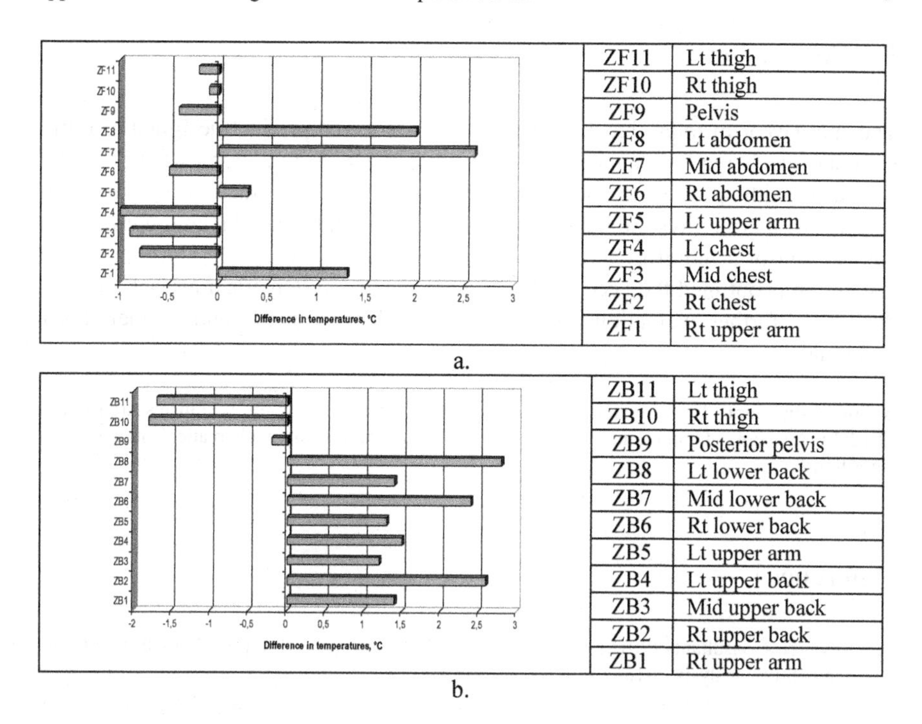

ZF11	Lt thigh
ZF10	Rt thigh
ZF9	Pelvis
ZF8	Lt abdomen
ZF7	Mid abdomen
ZF6	Rt abdomen
ZF5	Lt upper arm
ZF4	Lt chest
ZF3	Mid chest
ZF2	Rt chest
ZF1	Rt upper arm

a.

ZB11	Lt thigh
ZB10	Rt thigh
ZB9	Posterior pelvis
ZB8	Lt lower back
ZB7	Mid lower back
ZB6	Rt lower back
ZB5	Lt upper arm
ZB4	Lt upper back
ZB3	Mid upper back
ZB2	Rt upper back
ZB1	Rt upper arm

b.

Fig. 1 Changes of temperature on the surface of single layered sportswear: **a** frontal body zones, **b** back body zones

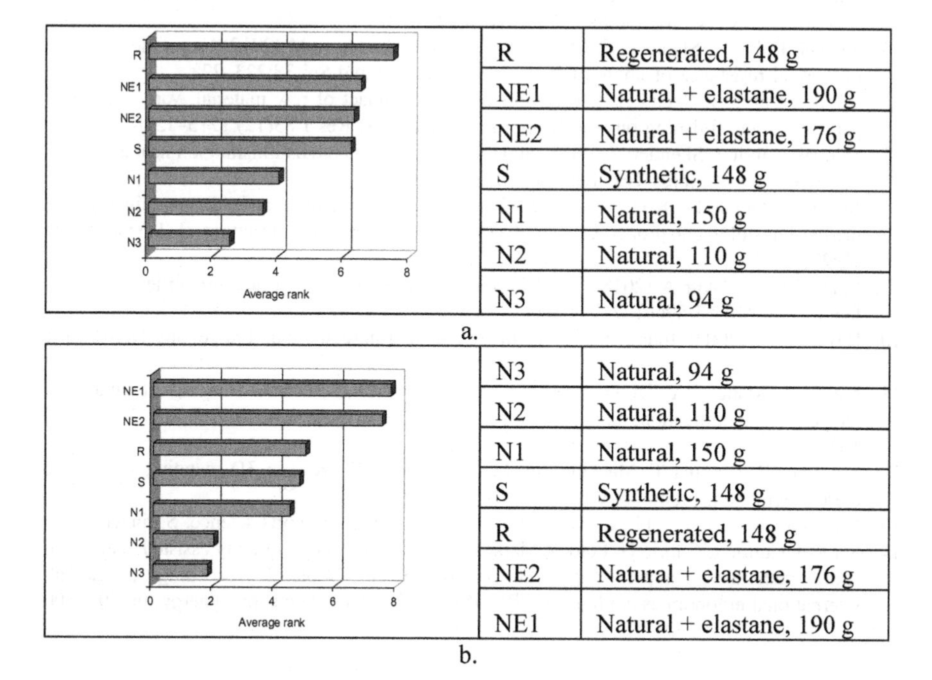

R	Regenerated, 148 g
NE1	Natural + elastane, 190 g
NE2	Natural + elastane, 176 g
S	Synthetic, 148 g
N1	Natural, 150 g
N2	Natural, 110 g
N3	Natural, 94 g

a.

N3	Natural, 94 g
N2	Natural, 110 g
N1	Natural, 150 g
S	Synthetic, 148 g
R	Regenerated, 148 g
NE2	Natural + elastane, 176 g
NE1	Natural + elastane, 190 g

b.

Fig. 2 Average ranks for smoothness (**a**) and softness (**b**)

5 Conclusion

The sportswear of today has become a truly engineered product designed to fulfill the consumer's requirements and to improve performance in sports what is tightly correlated with specifics of each sport. This paper covers evaluation of sportswear in terms of thermophysiological comfort of material and single layered garment, as well as sensorial comfort of single layered sportswear material. Throughout the discussion is outlined the need for comprehensive and systematic approach towards design and further evaluation of sportswear in order to enhance and augment body functions.

Acknowledgement This research is supported by University of Zagreb, within research grants TP12/20 "Design of functional sportswear with implication of mathematical models and algorithms".

References

1. Research Consultants (2018) Global sportswear industry research report, Growth trends and competitive analysis 2018–2025
2. Shishoo R (2015) Textiles for sportswear. Woodhead Publishing, Cambridge, United Kingdom
3. Ishtiaque SM, Sen K, Kumar A (2013) New approaches to engineer the yarn structure; part A: for better carpet performance. J Ind Text 44(4):605–624
4. Liu S, Zhou J (2019) Analysis of the dynamic properties of elastic knitted fabric for sportswear: inverse stress relaxation. Text Res J 89(9):1673–1683
5. Salopek Čubrić I, Skenderi Z, Mihelić Bogdanić A, Andrassy M (2012) Experimental study of thermal resistance of knitted fabrics. Exp Thermal Fluid Sci 38:223–228
6. Salopek Čubrić I, Skenderi Z, Havenith G (2013) Impact of raw material, yarn and fabric parameters and finishing on water vapor resistance. Text Res J 83(12):1215–1228
7. Salopek Čubrić I, Skenderi Z (2013) Impact of cellulose materials finishing on heat and water vapor resistance. Fibres Text East Eur 21(1):61–66
8. Jhanji Y, Gupta D, Kothari VK (2015) Effect of loop length and filament fineness on thermo-physiological properties of polyester-cotton plated knit structures. J Text Inst 106 (4):383–394
9. Balci Kilic G, Okur A (2018) Effect of yarn characteristics on surface properties of knitted fabrics. Text Res J 89(12):2476–2489
10. Havenith G (2003) Interaction of clothing and thermoregulation. Exogenous Dermatol 1 (5):221–230
11. Merla A, Mattei PA, Di Donato L, Romani GL (2010) Thermal imaging of cutaneous temperature modifications in runners during graded exercise. Ann Biomed Eng 38(1):158–163
12. Kinnicutt P, Domina T, MacGillivray M, Lerch T (2010) Knit-in 3D mapping's effect on thermoregulation: preliminary results. J Text Inst 101(2):120–127
13. Quesada JIP (2017) Application of infrared thermography in sports science. Springer
14. Salopek Čubrić I, Čubrić G, Potočić Matković VM (2016) Thermography assisted analysis of textile materials as a predictor of human comfort. In: Book of proceedings of the 6th international ergonomics conference ERGONOMICS 2016—Focus on synergy, pp 295–300

15. Salopek Čubrić I, Čubrić G, Perry P (2019) Assessment of knitted fabric smoothness and softness based on paired comparison. Fibers Polym. 20(3):656–667
16. Bertaux E, Lewandowski M, Derler S (2007) Relationship between friction and tactile properties for woven and knitted fabrics. Text Res J 77(6):387–396
17. Wiskott S, Weber MO, Haimlich F, Kyosev Y (2018) Effect of pattern elements of weft knitting on haptic preferences regarding winter garments. Text Res J 88(15):1689–1709

Functional and Subjective Assessment of Spinal Exoskeletons: From Development of Battery of Tests to Experiments with Low Back Pain Patients

Nejc Šarabon and Žiga Kozinc

Abstract Spinal exoskeletons have recently emerged as a potential tool to prevent and treat low back pain (LBP) in workers that are involved in heavy load handing and other demanding jobs. However, the widespread use of these devices is still limited. Notably, discomfort and hindrance of movement is often reported by the end-users. Therefore, there is an increasing interest to explore functional and subjective aspect of exoskeleton use. In this paper, we will present a novel battery of tests that was designed to comprehensively assess the effects of spinal exoskeletons on functional performance, discomfort and task difficulty. Next, we provide evidence that specific populations, such as LBP patients, may respond differently to the use of exoskeletons, which should be considered during design and development. Finally, we briefly summarize the findings from a study that investigated the effects of one of the state-of-the-art exoskeletons, by using the previously developed functional test battery.

Keywords Exosuit · Workload · Spine compression · Self-reported · Comfort · Low back pain

N. Šarabon (✉) · Ž. Kozinc
Faculty of Health Sciences, University of Primorska, Polje 42, 6310 Izola, Slovenia
e-mail: nejc.sarabon@fvz.upr.si

N. Šarabon · Ž. Kozinc
Andrej Marušič Institute, University of Primorska, Muzejski Trg 2, 6000 Koper, Slovenia

N. Šarabon
Human Health Department, InnoRenew CoE, Livade 6, 6310 Izola, Slovenia

N. Šarabon
Laboratory for Motor Control and Motor Behavior, S2P, Science to Practice Ltd, Tehnološki Park 19, 1000 Ljubljana, Slovenia

D. Sumpor et al. (eds.), *Proceedings of the 8th International Ergonomics Conference*, Advances in Intelligent Systems and Computing 1313,
https://doi.org/10.1007/978-3-030-66937-9_2

1 Introduction

Low back pain (LBP) is one of the major musculoskeletal conditions nowadays [1]. One of the possible solutions to battle LBP has emerged recently in the form of spinal exoskeletons (Fig. 1). These devices are designed to support prevention of LBP and facilitate the vocational reintegration of LBP patients in occupational environments that involve heavy-load handling and sustained awkward postures [2–4]. Exoskeletons provide external torque to the user, which is purported to reduce the required force from the muscles, and thereby also reduces the load on the spine and other joints. In recent years, numerous spinal exoskeletons have been biomechanically assessed [5–9]. Generally, the results from these studies confirm that the state-of-the-art exoskeletons can substantially reduce spinal loads, particularly during lifting tasks and certain static postures. Thereby, the basic concept of spinal exoskeleton for reduction of workload is well proven. However, before the end-users (workers) can truly benefit from the exoskeletons, they must be willing to use them.

It has been stressed that user-device interaction is an important barrier that prevents the state-of-the-art spinal exoskeletons to be more widely used [10]. In recent studies, it was relatively common for the participants to report discomfort associated with the exoskeleton [8, 11]. Moreover, it is important to acknowledge

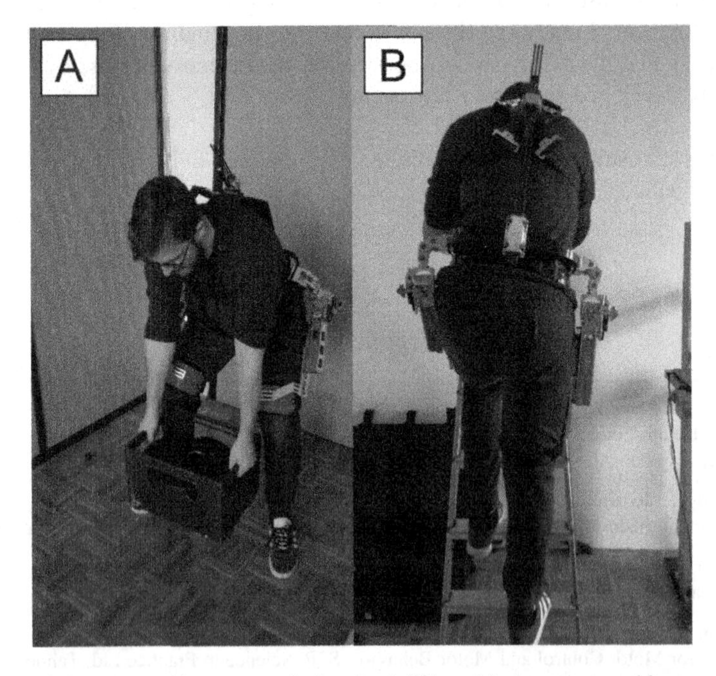

Fig. 1 Exoskeletons provide support during load lifting (**a**), however, problems may arise because the movement might be hindered during other tasks, such as ladder climbing (**b**)

that the majority of workplaces involve multiple tasks (Fig. 1b), not only lifting (Fig. 1a) or static postures. Therefore, the exoskeletons should also be able to accommodate such tasks that do not require support (e.g. walking, ladder climbing, etc.). The acceptance of the exoskeleton, regardless of the biomechanical support offered during strenuous tasks, will remain poor if the users fell uncomfortable during other movements and tasks. Therefore, it is critical that the end-user testing of the exoskeletons includes functional testing (i.e. assessing if the exoskeleton hiders freedom of movement during multitude of tasks), as well as the assessment of subjective perceptions (e.g. comfort, freedom of movement) and user impressions.

In this paper, we will present our recent endeavors to introduce functional and subjective testing of the exoskeletons to the overall battery of tests, which is predominantly comprised of biomechanical tests [5, 6, 9] and test of energy expenditure [7, 10]. First, we will describe a battery of test that was developed to comprehensively assess the functional performance and subjective responses of the exoskeleton users. Next, we will provide evidence for differences in subjective responses to exoskeleton use between LBP patients and asymptomatic individuals. Finally, we will summarize one of our studies that examined the effects of a passive spinal exoskeleton, in order to demonstrate the utility of the designed battery of tests.

2 Development and Validation of Battery of Functional and Subjective Tests

2.1 Background

One of the first evidence for potential limitations regarding the acceptance of the exoskeleton came from Baltrusch et al. [8]. Together with their research group, we designed a battery of tests, consisting of 12 functional tasks that are intended to comprehensively assess the functional performance during exoskeleton use. The selection of tasks was based on the list of tasks that are included in the functional capacity evaluation (FCE, Isernhagen Work Systems). This assessment method is used to realistically and reliably judge work-related physical performance capacity [12]. It involves postural holding tasks, lower lifting and 6-minute walk. Other tasks were added to the present test battery based on our own workplace observations.

2.2 The Battery of Tests and Study Design

To assess the reliability of the designed test battery, we invited twenty (10 females, 10 males) healthy participants for a repeated-measures study. The exclusion criteria were any current musculoskeletal injuries or any other medical conditions that could be exacerbated by the physical effort. The test battery was performed while wearing a novel SPEXOR passive spinal exoskeleton [2, 3] (Fig. 1).

The study was conducted during two sessions that were separated by 7–10 days. The first session included two sets of measurements (intra-session reliability). An additional single set was performed during the second session (inter-session reliability). The tasks were performed in a random order. The battery of tests includes tasks during which support of the exoskeleton is expected (Fig. 2a–d), tasks during which the exoskeleton can potentially hinder the performance (Fig. 2e–h) and tasks during which the exoskeleton can interfere with the range of motion (Fig. 2i–k). Additionally, a 6-min walking test was performed (not depicted on Fig. 2). The exoskeleton support was engaged only during tasks a–d. In others tasks, the support was turned off using the clutch. Namely, the SPEXOR exoskeleton enables quick and effortless engagement and disengagement of support by switching a clutch the hip module.

Most tasks are very short, while three tasks demand 5–6 min to be completed (two static postural tolerance tests and 6-minute walking). The outcome measures include objective (e.g. time to complete the task, distance covered, range of motion) and subjective (self-reported) performance measures. Subjective performance was assessed in terms of perceived task difficulty, perceived pain and discomfort. After each task, participants had to report (on a 0–10 visual analogue scale) regarding the abovementioned subjective aspects (0 indicating no discomfort/pain or no difficulties in performing the task, while 10 indicated highest discomfort or severe difficulties in performing the task). The discomfort rating referred strictly to the device (e.g. squeezing, pinching, rubbing of the device), while perceived task difficulty was assessed for task execution in general. For all tasks with time as an outcome measure, the instruction was to perform the task as fast as possible, but still in a safe manner.

Reliability was statistically analyzed two-way by mixed single intra-class correlation coefficients (ICC) with and standard error of measurement expressed as coefficients of variation (CV). For the assessment of the reliability of subjective outcomes, the Cronbach's α (internal consistency) and Spearman correlation coefficient were calculated. Pairwise t-tests and Wilcoxon tests were used to assess systematics bias.

2.3 Results and Conclusions

Generally, our results showed good reliability of the outcomes (ICC > 0.75; CV < 10%). Interestingly, load lifting was consistently the most problematic task, both for the objective outcome (number of lifts; ICC = 0.49–0.74), and subjective outcomes (α = 0.58–0.69). Systematic bias was present between the sessions for several tasks regarding discomfort rates (load lifting, load carrying, 6-minute walk, squatting and ladder climbing). In all cases, the discomfort was lower, indicating

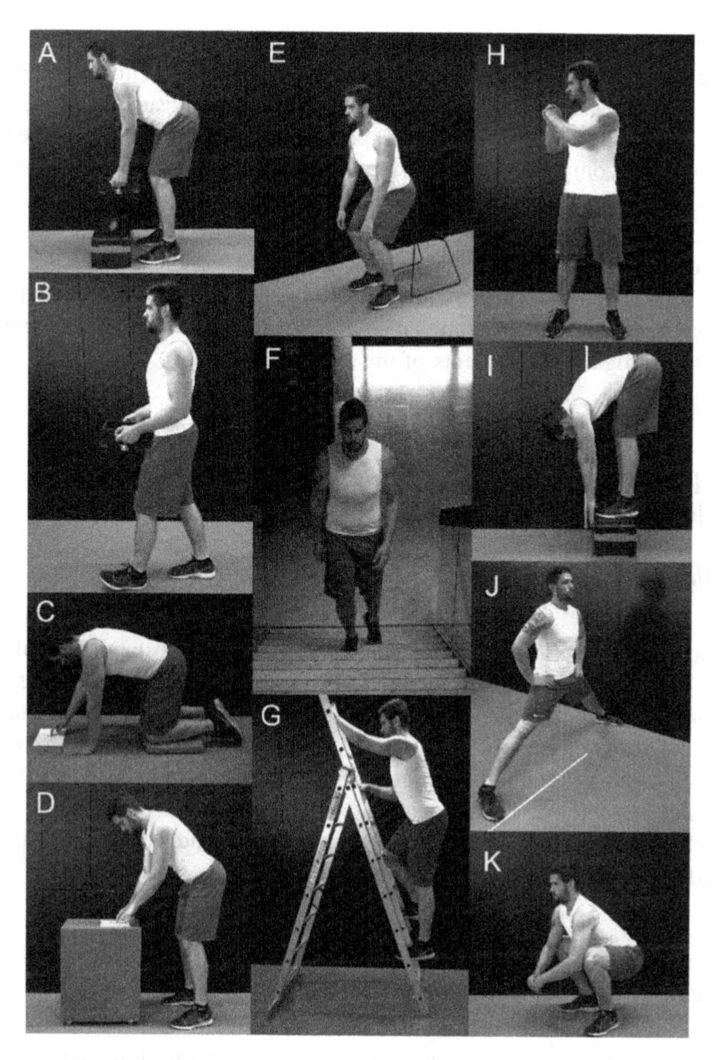

Fig. 2 The tasks included in the test battery. Additionally, a 6-min walking test was performed (not depicted)

that the participants got used to the exoskeleton. Subjective outcomes tended in general toward lower discomfort or lower task difficulty during later trials, which could indicate that the tolerance of the device increases with time of use. However, it remains unknown how the scores would change with longer-use (e.g. though a work shift.)

Nevertheless, most of the outcome measures of the presented battery of tests showed good to excellent reliability. The presence of systematic errors and lower ICC scores in some tasks warrants caution when examining the effects of exoskeleton using the present battery of tests. Because the reliability is lower inter-session, particular caution is needed when assessing the long-term effects. Providing more practice trial tasks is a potential solution to improve reliability, however, further research is needed to investigate how many such trials are needed. Moreover, it needs to be stressed that the present battery of tests is not to be used as a single tool for comparison of exoskeletons. The evaluation of the exoskeletons should include other testing approaches, such as detailed biomechanical testing [5, 6] and physiological testing (e.g. monitoring metabolic cost when wearing the exoskeleton [7, 10]). The present battery of tests represents an easy-to-use tool that could be used to provide important feedback to the exoskeleton developers and could represent an important complementary tool to the conventional biomechanical and physiological assessments of the exoskeletons.

3 Responses to Exoskeleton Use in Low Back Pain Patients

3.1 Objective and Methods

While the state-of-the-art exoskeletons were shown to substantially unload the back, the user acceptance is still limited [13]. Perceived discomfort and restriction of freedom of movement are commonly reported. In this small pilot study, we explored the differences in subjective responses and user impressions to using passive spinal exoskeleton during a set of simple lifting tasks between LBP patients ($n = 12$) and asymptomatic individuals ($n = 10$). Participants performed a series of lifting tasks while wearing the SPEXOR exoskeleton, and visual analogue scales were used for all assessments of subjective responses.

3.2 Results and Conclusions

Overall, the results showed mostly similar responses in healthy individuals and LBP patients. In some tests, the responses were slightly more positive for LBP patients. Most notably, the LBP patients reported a statistically significantly ($p = 0.048$) higher willingness to use the device in everyday life (5.36 ± 4.05) compared to the control group (2.00 ± 1.85) and also gave the device a higher overall grade (6.58 ± 1.98 vs. 4.30 ± 2.26; $p = 0.021$).

Therefore, this study has demonstrated that individuals with present LBP problems responded differently (i.e. more favorably) to the use of the spinal exoskeleton for simple lifting tasks. This implies that current exoskeletons could be appropriate for LBP rehabilitation, but not prevention, as healthy individuals are less willing to use such devices. Future studies should explore whether different exoskeleton designs could be more appropriate for people with no LBP issues.

4 Investigation of Exoskeleton Effect on Functional Performance and Subjective Responses in Patients with Low Back Pain

4.1 Objectives and Rationale

After the test battery was completed and methodologically tested, our attention was turned to exploring the effects of the novel SPEXOR passive exoskeleton in LBP patients. The design of the SPEXOR exoskeleton (Fig. 1) is explained in detail by Näf et al. [3]. Most notably, the novelty of the SPEXOR lies the possibility to disengage the support by a manual clutch, and the inclusion of a self-aligning mechanism at the hip joint, which improves the fitting to each individual. Biomechanical studies have shown that the SPEXOR importantly reduces the workload and spine compression [5, 6]. This suggests that the SPEXOR is useful for reducing the risk of LBP in workers who perform lifting tasks or spend prolonged time in static postures, such as forward bend. Moreover, the SPEXOR could be used in rehabilitation and reintegration of patients that already suffer from LBP. However, as we have seen in the previous section, the effects of wearable robotic devices are sometimes specific to the studied population. Therefore, LBP patients could respond differently in light of the scores in our test battery. The purpose of our final study was to assess how the SPEXOR affects functional performance and discomfort in patients with current LBP.

4.2 Methods

We used the battery of test that we have previously developed. The participants performed the test with or without the SPEXOR exoskeleton (in a random order). We tested 14 participants (7 females, 7 males; age 40.5 ± 10.8 years) with current moderate LBP (2–7 on a 0–10 scale). All measurements were performed within a single session lasting ~90 min. All of the procedures (tasks, outcome measures and subjective response assessment are) are completely the same as in the reliability study. For assessing differences in objective measures, 2-tailed paired t-test was used. For differences in subjective measures, we used Wilcoxon signed-rank test.

4.3 Results

Regarding the objective outcomes, there was a statistically significant improvement in mean time of holding the posture in static forward bending task (Fig. 1d) (with exoskeleton: 227.1 ± 84.6 s; without exoskeleton: 255.0 ± 65.3 s; $p = 0.009$). On the other hand, we recorded negative effects of the SPEXOR on time for sit-stand test (Fig. 1e) (with exoskeleton: 12.7 ± 4.4 s; without exoskeleton: 11.6 ± 4.5 s; p = 0.009) and ladder climbing (Fig. 1g) (with exoskeleton: 17.9 ± 5.5 s; without exoskeleton: 16.6 ± 5.3 s; p = 0.009). No other effects of the exoskeleton were found.

Regarding the perceived task difficulty, the only significant difference between conditions was found for Sit-stand task (Fig. 1e) for which the SPEXOR decreased difficulty scores according to Wilcoxon signed rank test (p = 0.04), although the median values were the same (0.3). There was no clear trend indicating that exoskeleton affects perceived task difficulty.

There was a tendency for participants to report lower low back discomfort scores when wearing the exoskeleton, however, statistically significant differences were found only for static forward bending (difference of medians = 0.8; p = 0.028) and Sit-stand tasks (difference of medians = 1.3; p = 0.013).

The participants rated the device as being easy to adjust and somewhat difficult to put on and off. They reported that the device did not restrict their movement considerably or interfere with the tasks that they performed. They assessed the support of the device mostly as low to modest.

5 Conclusion

The use of spinal exoskeletons is still limited, despite several experiments confirming their positive effect on spinal load and energy demands for work. This is largely attributable to the robust design of the exoskeletons that causes discomfort, or hinders the movement during the tasks that do not require support. Recent studies, some of which have been summarized in this paper, have started to address this issue. However, future research and development will be needed to design exoskeleton models that will be well accepted by the end-users.

References

1. Foster NE, Anema JR, Cherkin D, Chou R, Cohen SP, Gross DP et al (2018) Prevention and treatment of low back pain: evidence, challenges, and promising directions. Lancet 391: 2368–2383
2. Babič J, Kingma I, Bornmann J, Mombaur K, Näf M, Petrič T et al (2019) SPEXOR: design and development of passive spinal exoskeletal robot for low back pain prevention and vocational reintegration. SN Appl Sci 1:454
3. Näf MB, Koopman AS, Baltrusch S, Rodriguez-Guerrero C, Vanderborght B, Lefeber D (2018) Passive back support exoskeleton improves range of motion using flexible beams. Front Robot AI 5
4. de Looze MP, Bosch T, Krause F, Stadler KS, O'Sullivan LW (2016) Exoskeletons for industrial application and their potential effects on physical work load. Ergonomics 59: 671–681

5. Koopman AS, Kingma I, Faber GS, de Looze MP, van Dieën JH (2019) Effects of a passive exoskeleton on the mechanical loading of the low back in static holding tasks. J Biomech 83:97–103
6. Koopman AS, Kingma I, de Looze MP, van Dieën JH (2020) Effects of a passive back exoskeleton on the mechanical loading of the low-back during symmetric lifting. J Biomech 102
7. Baltrusch SJ, van Dieën JH, Koopman AS, Näf MB, Rodriguez-Guerrero C, Babič J et al (2020) SPEXOR passive spinal exoskeleton decreases metabolic cost during symmetric repetitive lifting. Eur J Appl Physiol 120:401–412
8. Baltrusch SJ, van Dieën JH, van Bennekom CAM, Houdijk H (2018) The effect of a passive trunk exoskeleton on functional performance in healthy individuals. Appl Ergon 72: 94–106
9. Bosch T, van Eck J, Knitel K, de Looze M (2016) The effects of a passive exoskeleton on muscle activity, discomfort and endurance time in forward bending work. Appl Ergon 54:212–217
10. Baltrusch SJ, van Dieën JH, Bruijn SM, Koopman AS, van Bennekom CAM, Houdijk H (2019) The effect of a passive trunk exoskeleton on metabolic costs during lifting and walking. Ergonomics 62:903–916
11. Kozinc Ž, Baltrusch S, Houdijk H, Šarabon N (2020) Short-term effects of a passive spinal exoskeleton on functional performance, discomfort and user satisfaction in patients with low back pain. J Occup Rehabil 2020 [online ahead of print]
12. Reneman MF, Brouwer S, Meinema A, Dijkstra PU, Geertzen JHB, Groothoff JW (2004) Test-retest reliability of the Isernhagen work systems functional capacity evaluation in healthy adults. J Occup Rehabil 14:295–305
13. de Looze MP, Krause F, O'Sullivan LW (2017) The potential and acceptance of exoskeletons in industry. Biosyst Biorobotics 16:195–199

Session Papers

Skidder Operator Vibration Exposure

M. Šušnjar, Z. Pandur, Z. Zečić, H. Kopseak, and M. Bačić

Abstract The exposure of workers to vibration is expressed as energy equivalent A(8), which is determined by the procedure clearly described in the international standard ISO 5349-1-2001. A(8) is a value that depends not only on the vibration level in certain operating procedures, but also on the duration of exposure. Research was conducted on Ecotrac 120 V and Ecotrac 140 V, the most used skidders in Croatian forestry. Vibration on the steering wheel and seat of the skidders was measured by vibrometer with triaxial accelerometer. Measurements were performed at three characteristic engine operating modes: idling, at maximum torque and full throttle. Daily exposure to vibration of a skidder operator A(8) is calculated based on structural data of daily working times in different working conditions as well as distribution of three characteristic engine operating modes of effective time. The results show differences in vibration levels between different types of tractors and working conditions. The research was carried out within the project "Development of hybrid skidder—HiSkid" co-financed by European Regional Fund in the scope of European Union Operational Program "Competitiveness and Cohesion" 2014–2020 (2014HR16M1OP00-1.2).

Keywords Skidder · Daily vibration exposure · A(8) · WBV · HAV · Engine operating modes

M. Šušnjar · Z. Pandur · Z. Zečić · H. Kopseak · M. Bačić (✉)
Faculty of Forestry, University of Zagreb, Zagreb, Croatia
e-mail: mbacic1@sumfak.hr

M. Šušnjar
e-mail: msusnjar@sumfak.hr

Z. Pandur
e-mail: zpandur@sumfak.hr

Z. Zečić
e-mail: zzecic@sumfak.hr

H. Kopseak
e-mail: hkopseak@sumfak.hr

D. Sumpor et al. (eds.), *Proceedings of the 8th International Ergonomics Conference*, Advances in Intelligent Systems and Computing 1313,
https://doi.org/10.1007/978-3-030-66937-9_3

1 Introduction

The ergonomic aspects of forestry machines with respect to operator exposure to machine vibrations have become the primary consideration for enhancing operator performance in recent decades. In mechanized forest operations, whole-body vibration may have considerable influence on machine productivity and the health of the operator [1]. Whole-body vibration is a problem with forestry machines and often causes discomfort and injury. The Directive 2002/44/EC [2] defines the minimum health and protection requirements for the workers exposed to vibrations that are transmitted to the hand-arm system (HAV—*hand-arm vibrations*) and to the whole body of the operator (WBV—*whole-body vibrations*). Directive 2002/44/EC unambiguously sets daily exposure action value of 2.5 m/s^2 (HAV) and 0.5 m/s^2 (WBV) as a warning level above which protective measures must be applied and daily exposure limit value of 5 m/s^2 (HAV) and 1.15 m/s^2 (WBV) above which work must be stopped. The A(8) value does not only depend on the vibration level in certain operating procedures, but also on the duration of exposure, i.e. on the duration of each skidder operating procedure [3]. In the forestry field, previous studies have shown that forestry vehicle operators are exposed to high levels of WBV. Limit values could be exceeded with a large number of modern farming tractors [4]. One study measured operator exposure to whole-body vibration in timber extraction with grapple skidder when comparing the values of the weighted acceleration and VDV (*vibration dose value*) with the limits established by the European Directive, it is verified that the measured values are higher than the one expected by the respective regulation [5]. Factors such as vehicle speed, driving style, machine design and geometry, belt and chain type, suspension type, power transmission to the wheels, mass distribution, position of the driver's seat, and the seat features affect the exposure of operators of forestry machinery to WBV [6]. Measured vibrations on steering handles of the one-axle farming tractor during driving and soil processing at three tractor speeds indicate the highest vibrations at the lowest tractor speed and during soil processing, and actually higher vibrations were measured during soil processing [7]. Worker's exposure to whole-body vibration is the highest during the skidding operations and for operations where a cable skidder was moving with no load (1.31 m/s^2) and ramping (1.22 m/s^2), and the lowest with the full load (0.91 m/s^2) [8]. These exposure levels increase with increases in driving speed, the roughness of the terrain, and decreases in driving with a load, as seen in forestry skidders [9]. In harvesters the vibration level is affected by the characteristics of the vehicle (engine speed, engine fitted with shock-absorbers), terrain characteristics (surface obstacles), methods of wood processing (processing of trees), seat characteristics, physical condition and sitting position of the operator and soil characteristics (dry, frozen, wet soil). The same authors conclude that the air pressure in tires has a considerable impact on the level of vibrations that are transmitted through the seat to the whole body of the operator while the harvester is moving on uneven terrain [10]. Vehicle speed and the type of terrain on which it moves highly affect the level of vibrations that are transmitted through the seat to the whole body of the operator [11]. Technical factors that could be used to help minimize operator WBV exposures such as: seat suspension systems, seat cushioning, cab/chassis, and axel/chassis suspension systems, and the vehicle tires inflation pressure are presented in one study [12]. Higher WBV operator exposures are detected during tractor skidding than the limit values specified by international standards. But when a skidding tractor has a seat suspension system with springs, significant reductions is noticed in the total WBV accelerations [1]. Aside from technical upgrades of machines, reduction of exposure to WBV, while simultaneously maintaining high productivity, requires careful selection of worksites and adapted work organization [13]. With the methodology of determining 8-hour energy equivalent of the total value of the estimated accelerations A(8), an accurate picture of the working day should first be made and whole day shooting of the operator's work with the film camera is one of the ways to get such a picture. The same authors state that, in practice, it is practically impossible to measure the levels of vibrations for each activity, and for this reason, it is necessary to make initial measurements in the test polygon under controlled conditions [14]. The vibration exposures measured on these test tracks do not necessarily represent actual field exposures. Exposures measured on ISO 5008 test tracks overestimate those found during field tasks [15]. On the other hand, one should be aware that studies tracking exposures over an entire work cycle may underestimate the acceleration levels for specific work tasks. This is due to the different vibration exposures for the individual work tasks as well as the fact that periods, where the vehicle is stationary, are often incorporated into the exposure average [16]. The aim of the research is to determinate the level of vibration of a skidder in controlled test conditions, and modeling the total value of the exposure to vibration of a skidder driver on a daily basis A(8). The research results are used to investigate vibration exposure in real working conditions and represent the baseline values for future future research within the project "Development of hybrid skidder —HiSkid".

2 Materials and Methods

The research was done on the skidders Ecotrac 120 V and Ecotrac 140 V, which are the most used skidders in Croatian forestry. The measurements were carried out on the skidders Ecotrac 120 V, whose mass is approximately 7.5 tons, and on the Ecotrac 140 V, whose mass is approximately 8 tons. Ecotrac 120 V is powered by a 6-cylinder air-cooled engine with a nominal power of 86 kW and the Ecotrac 140 V is powered by a 4-cylinder liquid-cooled engine with a nominal power of 104 kW. The skidders are fitted with an air suspension seat whose sensitivity can be regulated manually. For the data obtained from measurements in three modes at all measuring points, the triaxial vibrometer calculated the weighted vibration values for time records in the direction of all three axes, in accordance with the recommendations of the standard HRN EN ISO 5349:2008 [17, 18]. Vibration on the steering wheel and seat of the skidders was measured using vibrometer Brüel & Kjaer type 4447 and triaxial accelerometer type 4520-002 with UA 3017 mount on the steering wheel and triaxial accelerometer type 4524-B fitted in the rubber protective cover on the seat of the researched skidders (Fig. 1). Measurements were performed at three characteristic engine operating modes: idling (900 rpm), at maximum torque (1500 rpm), and full throttle (2200 rpm). Weighted acceleration levels in all three axes (a_{hwx}, a_{hwy} and a_{hwz}) were obtained. From the weighted acceleration levels in all three axes, the vibration total value (a_{hv}) for three modes according to the following relation was determined (Eq. 1):

$$a_{hv} = \sqrt{a_{hwx}^2 \cdot a_{hwy}^2 \cdot a_{hwz}^2} \quad (1)$$

a_{hwi}—weighted effective acceleration levels (m/s^2).

The exposure to vibration of a skidder driver on a daily basis A(8) is calculated based on data of the structure daily working times (effective time and allowance time) in different working conditions (hilly terrain for Ecotrac 120 V, and mountainous terrain for Ecotrac 140 V) as well as the distribution of three characteristic engine operating modes of effective time. The mentioned data are taken from previous scientific research on the work of skidders [19]. Based on the measured acceleration values on the seat and steering wheel, and the calculated vibration total values, a model for estimating the daily exposure value A(8) was calculated according to the relation (Eq. 2):

$$A(8) = \sqrt{\frac{1}{T_0}\sum_{i=1}^{N} a_{hvi}^2 \cdot T_i} \quad (2)$$

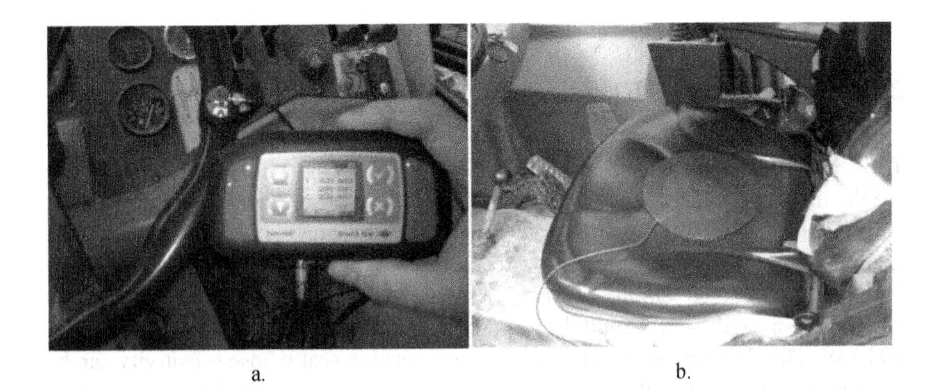

a. b.

Fig. 1 **a** Measuring of vibrations on the steering wheel. **b** Measuring of vibrations on the seat

Table 1 Structure of daily working times

		%	min
ECOTRAC 120 V hilly terrain	Total effective time	69	330
	<1200 rpm	6.9	33
	1200–1800 rpm	55.2	265
	>1800 rpm	6.9	33
	Allowance time	31	150
	Total time	100	480
ECOTRAC 140 V mountainous terrain	Total effective time	78	374
	<1200 rpm	7.8	38
	1200–1800 rpm	62.4	298
	>1800 rpm	7.8	38
	Allowance time	22	106
	Total time	100	480

T_0 daily working time of 8 h or 28,800 s
a_{hvi} vibration total value for i operation
T_i duration of i operation
N total number of operations

Additional time represents the non-productive part of the total working time related to preparatory time, mealtime, resting time, and other interruptions. Research on the productivity of Ecotrac 120 skidder determined an additional time factor of 1.22 for a site on mountain terrain, and a factor of 1.31 for a site on hilly terrain (Table 1) [19]. Based on these data, it is concluded that the additional time does not depend so much on stand and exploitation factors as on the organization and supervision of the execution of work in skidding wood. For the same skidder, in another study, in the hilly conditions of central Croatia the additional time factor is 1.34, and in the mountainous conditions of selective forests 1.18 [20]. In research [3], engine speed data are sorted during skidder normal work in the 2 working day period (Table 2). All data of engine speeds were obtained by the use of the FMS (*fleet management system*). All engine speeds less than 1200 rpm represent idle operating mode, engine speeds from 1200 to 1800 rpm represent normal operating mode near the maximum torque, and engine speeds higher than 1800 rpm represent full throttle operating mode.

Table 2 Daily vibration exposure

ECOTRAC 120 V	*Seat—WBV*	a_{hv} (m/s^2)	T (min)	A(8)
	Off	0	149	0.65
	Idle (900 rpm)	1.17	33	
	Max torque (1500 rpm)	0.74	265	
	Full throttle (2200 rpm)	0.65	33	
	Total		480	
	Steering wheel—HAV	a_{hv} (m/s^2)	T (min)	A(8)
	Off	0	149	0.73
	Idle (900 rpm)	0.63	33	
	Max torque (1500 rpm)	0.84	265	
	Full throttle (2200 rpm)	1.27	33	
	Total		480	
ECOTRAC 140 V	*Seat—WBV*	a_{hv} (m/s^2)	T (min)	A(8)
	Off	0	106	1.12
	Idle (900 rpm)	0.89	38	
	Max torque (1500 rpm)	1.38	298	
	Full throttle (2200 rpm)	0.24	38	
	Total		480	
	Steering wheel—HAV	a_{hv} (m/s^2)	T (min)	A(8)
	Off	0	106	1.23
	Idle (900 rpm)	1.13	38	
	Max torque (1500 rpm)	1.32	298	
	Full throttle (2200 rpm)	2.05	38	
	Total		480	

3 Research Results

Obtained results show that HAV values measured on the steering wheel do not even exceed the action value of 2.5 m/s^2. However, WBV values on the seat exceed action value in both observed skidders. Especially concerning is how close is Ecotrac 140 to exceeding the daily limit value for WBV of 1.15 m/s^2. The reason for that could be the different propulsion engine in Ecotrac 140. Ecotrac 140 and Ecotrac 120 are very similar in build, and both have pneumatic seats, so the reason for higher vibrations on the seat of the Ecotrac 140 could be the liquid-cooled, 4-cylinder, turbo diesel engine. On the other hand, Ecotrac 120 is propelled with an air-cooled, 6-cylinder, naturally aspirated diesel engine which is perhaps more balanced, and thus produces fewer vibrations.

4 Conclusions

Both of the observed skidders are designed very similar and with the same intention, to skid wood. Since the vibration exposure was measured in a controlled environment, and obtained vibration values do not take into account the terrain characteristics and other working factors, the main reason for the significant difference in WBV values is the propulsion engine. Winching makes a significant amount of work time in skidding during which the engine must operate in order to propel the winch. If the skidder would be fitted with batteries and electric powered winch, the total engine operating time would be reduced, and thus WBV would be lower. Electric motors are far more balanced than ICE (*internal combustion engine*) and also produce fewer vibrations. Regarding the latter, hybrid technology is already used in forestry and it is the logical step to ensuring better results in many fields including work ergonomics.

Acknowledgements The research was carried out within the project "Development of hybrid skidder—HiSkid" co-financed by European Regional Fund in the scope of European Union Operational Program "Competitiveness and Cohesion" 2014–2020 (2014HR16M1OP00-1.2).

References

1. Melemez K, Tunay M, Emir T (2013) The role of seat suspension in whole-body vibration affecting skidding tractor operators. J Food Agric Environ 11(2):1211–1215
2. Directive 2002/44/EC of European parliament and of the council: the minimum health requirement regarding to exposure of workers to the risks arising from physical agents (vibration). Offical Journal of the European Communities 177, pp 13–19
3. Pandur Z, Horvat D, Šušnjar M, Zorić M (2013) Possibility of determination of daily exposure to vibration of skidder drivers using fleet manager system. Croatian J For Eng 34(2): 305–310
4. Scarlett AJ, Price JS, Stayner RM (2002) Whole-body vibration: initial evaluation of emissions originating from modern agricultural tractors. Health and Safety Executive Books, pp 1–26
5. Cazani AC, Miyajima RH, Simões D, Santos JEG (2020) Operator exposure to whole-body vibration in timber extraction with grapple skidder. J Vib Eng Technol. https://doi.org/10.1007/s42417-020-00219-0
6. Tiemessen IJ, Hulshof CTJ, Frings-Dresen MHW (2007) An overview of strategies to reduce whole-body vibration exposure on drivers: a systematic review. Int J Ind Ergon 37(3): 245–256
7. Dewangan KN, Tewari VK (2009) Characteristics of hand-transmitted vibration of a hand tractor used in three operational modes. Int J Ind Ergon 39:239–245
8. Poje A (2011) Vplivi delovnega okolja na obremenitev in težavnost dela sekača pri različnih organizacijskih oblikah. Dokt. disertacija. Ljubljana, Univ. v Lj., Biotehniška fakulteta, Odd za gozdarstvo in gozdne vire, 1–232
9. Cation S, Jack R, Oliver M, Dickey JP, Lee-Shee NK (2008) Six degree of freedom whole-body vibration exposure during forestry skidder operations. Int J Ind Ergon 38: 739–757

10. Sherwin LM, Owende PMO, Kanali CL, Lyons J, Ward SM (2004) Influence of tyre inflation pressure on whole-body vibrations transmitted to the operator in a cut-to-length timber harvester. Appl Ergon 35(2004):253–261
11. Kumar S (2004) Vibration in operating heavy haul trucks in overburden mining. Appl Ergon 35:509–520
12. Jack RJ, Oliver M (2008) A review of factors influencing whole-body vibration injuries in forestry mobile machine operators. Int J For Eng 19(1):51–65
13. Poje A, Grigolato S, Potocnik I (2019) Operator exposure to noise and whole-body vibration in a fully mechanised CTL forest harvesting system in Karst terrain. Croatian J For Eng 40:139–150
14. Goglia V, Suchomel J, Žgela J, Đukić I (2012) Izloženost vibracijama šumarskih radnika u svjetlu Directive 2002/44/EC. Šumarski list, CXXXVI(5–6):283–289
15. Scarlet AJ, Price JS, Stayner RM (2007) Whole body vibration: evaluation of emission and exposure levels arising from agricultural tractors. J Terrramech 44:65–73
16. Jack RJ, Oliver M, Dickey JP, Cation S, Hayward G, Lee-Shee N (2010) Six-degree-of-freedom whole-body vibration exposure levels during routine skidder operations. Ergonomics 53(5):696–715
17. HRN EN ISO 5349-1:2008. Mechanical vibration—measurement and evaluation of human exposure to hand transmitted vibration. In: Part 1: general requirements. International Standard Organization, Geneva
18. HRN EN ISO 5349-2:2008. Mechanical vibration—measurement and evaluation of human exposure to hand transmitted vibration. In: Part 2: practical guidance for measurement at the workplace. International Standard Organization, Geneva
19. Zečić Ž, Benković Z, Papa I, Marenče J, Vusić D (2019) Productivity of tractor Ecotrac 120 V timber skidding in hilly area of central Croatia. Nova mehanizacija šumarstava 40:1–10
20. Horvat D, Zečić Ž, Šušnjar M (2007) Morphological characteristics and productivity of skidder ECOTRAC 120V. Croatian J For Eng 28(1):11–25

A Preliminary Study of the Possible Impact of Non-related Driving Tasks on the Car Driver's Performance Using Glance-Based Measures

Eva Bolčević, Sandro Tokić, Davor Sumpor, and Jasna Leder Horina

Abstract In road traffic, the knowledge of the impact of drivers' distraction on the performance is critical to address the safety incidents and to reduce the number of road accidents. Previous research has mostly investigated the impact of different types of driver distraction on the driver's performance. This preliminary research will explore the possible impact of simultaneous interaction of two or more types of distraction inside a vehicle on the driver's performance using the driver's glance behavior. A possible impact on the driver's performance will be analysed by comparing the driver's gaze parameters for different driver activities imposed by the scenario in relation to the baseline driving mode. The drivers will perform real-world driving scenarios doing a non-related driving task such as smoking, consuming soft drinks, and using a mobile phone fixed on a stand under the rearview mirror. Two scenarios have been used to evaluate mobile phone use: casual video call conversation and video call conversation while solving a cognitive task. The drivers' glance parameters such as glance duration, gaze direction, percent of gaze on road center, and others will be assessed from video data on the sample of three drivers. This preliminary study will be a small part of a larger-scale research planned for the near future.

Keywords Driver distraction · Mobile phone · Naturalistic driving · Driver

E. Bolčević · S. Tokić · D. Sumpor (✉)
Faculty of Transport and Traffic Sciences, University of Zagreb, Vukelićeva 4, 10000 Zagreb, Croatia
e-mail: dsumpor@fpz.unizg.hr

E. Bolčević
e-mail: evabolcevic@gmail.com

J. L. Horina
Faculty of Mechanical Engineering and Naval Architecture, University of Zagreb, Ivana Lučića 5, 10002 Zagreb, Croatia

D. Sumpor et al. (eds.), *Proceedings of the 8th International Ergonomics Conference*, Advances in Intelligent Systems and Computing 1313,
https://doi.org/10.1007/978-3-030-66937-9_4

1 Introduction

In recent years, in the field of driver distraction research, the assessment of the driver's field of view has become a popular topic because it is considered that the driver's focus is an essential characteristic of the driver's behavior and their attention to the task of driving [1]. For assessing the driver's field of vision, head and eye pose can be used, defined in three degrees of freedom (yaw, pitch, and roll). In real conditions, the drivers use head and eye movements when looking in certain zones [2]. There is a difference in how amateurs and professionals look. Amateurs usually make glances by moving their eyes without moving their head, while experienced drivers move their head and eyes simultaneously [3]. Therefore, it is best to make an assessment based on the position of the eyes and the position of the head.

The experiment took place at the Borongaj Science Campus (ZUK Borongaj) in Zagreb. The participants were filmed with high-quality cameras. There are two main goals of the research conducted on a pilot sample of three male student drivers who were driving according to the scenario shown in Fig. 1 and in Table 1. The first one is to master the remote-sensing technologies with the help of a camera installed in the vehicle cab and try to bypass one of the system main drawbacks, which is sensitivity to external condition lighting [4]. The second one is testing and improving the research design to assess the factors of cabin distractions of road vehicle drivers using the driver's glance behavior method on a larger number of participants.

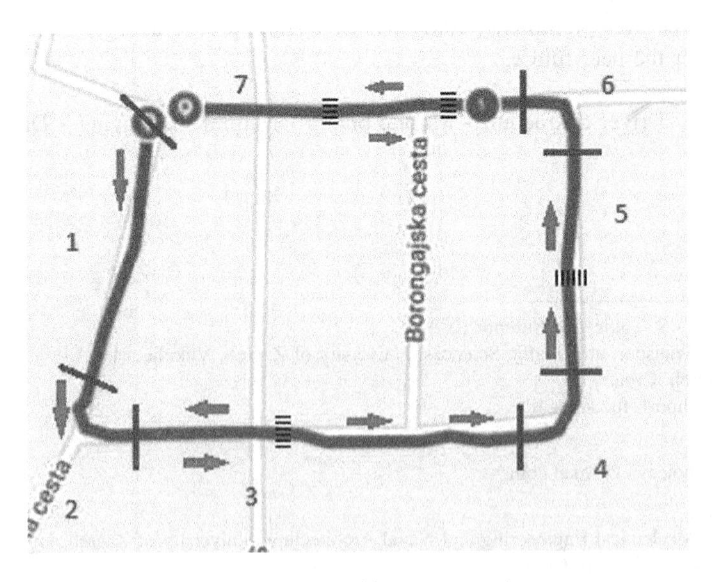

Fig. 1 Separation of the route in ZUK Borongaj into seven sections

Table 1 Separation of the route at ZUK Borongaj into seven sections

Section label	Distance L_s (m)		Length L (m)	Short name	Full name of the section
	Initial	Final			
1	000	250	250	Flat part west	Fully flat one-way section
2	250	300	50	Sharp left	Shart left turn + intersection
3	300	580	280	Flat part south	Fully flat two-way and one-way section
4	580	630	50	Sharp left	Shart left turn
5	630	830	200	Flat part east	Fully flat one-way section
6	820	870	50	Shart left	Sharp left turn + intersection
7	870	1120	250	Flat part north	Fully flat two-way section

The road drivers' performance in complex traffic and working environment in baseline driving contains more common and often simultaneous activities while maintaining direction and speed, which without additional distractions can significantly impair the driver performance and cause loss of control over the vehicle. In this research, the so-called baseline driving means everyday driving that contains all the drivers' usual activities, without secondary distraction tasks (visual, cognitive, manual or a combination of the above). To detect and evaluate the factors of cabin distractions of passenger vehicle drivers by using the method of driver gaze analysis, it is first necessary to determine the average values and limit ranges of the amount of driver glance parameters for baseline driving. Eye movement measures such as glance duration, glance frequency (number of glances in zones, and number of used zones), percent of gaze on road center (specifically to the main zone, the "Left part of the windshield"), and others, will be analyzed from the video data on the sample of three drivers.

2 Research Draft

In addition to the driver, three other persons participate in the research. The first one is a passenger next to the driver that gives participant instructions according to the scenario and turns on/off the cameras and the Endomondo platform. The second passenger in the back seat monitors the scenario implementation and gives a signal to the assistant. The assistant outside the car performs activities B3 and D1, which involve the use of video connection in the Viber application on a mobile phone.

The driver's glance behavior was analyzed based on eleven possible driver view zones:

(1) Left side of the windshield;
(2) Left side mirror;
(3) Rear-view mirror;
(4) Right side mirror;
(5) Instrument panel;
(6) Navigation (central part of the windshield);
(7) Center Console;
(8) Left window;
(9) Right window;
10) Others (beneath the steering wheel).

It is important to note that for driver activities (A), (B3) and (D1) the cell phone is placed on the central part of the windshield, on the stand on the left and under the rearview mirror, within the "Navigation" zone, for the following reasons:

(a) to minimize the impact of manual distraction due to holding the mobile phone with the dominant hand during the above activities;
(b) because it is now a common situation in real traffic (drivers use mobile phones over a wireless connection with an earpiece, using headphones without accessories fixed on a stand), and because legal regulations in the Republic of Croatia require drivers to keep both hands on to the steering wheel while driving;
(c) research [5] confirms that the effect of using a mobile phone while driving on the performance of the driver is the same regardless of whether the driver holds the mobile phone with his hand during the conversation or not.

The driver's activities specified by the scenario to cause cabin's distraction are as follows:

(A) **Setting up navigation on a cell phone**: location and driving route setting by using Google maps application (n = 1, average duration 45 s).
(B3) **Mobile phone conversation, video call**: talk to assistant by using Viber application (n = 2, duration in the range 55–109 s).
(D1) **Cognitive task through mobile phone conversation, video call**: (n = 2, duration in the range 102–114 s). Two-back task implies that the assistant reads a series of letters one by one every 3 s, and the participant needs to say "same" when he hears the same letter that is identical to the letter two positions before.
(E) **Complex task** (n = 1, average duration 101 s), and to serve multiple commands on the center console:

  (a) Adjusting radio (turn on the radio, change few radio stations, adjust the sound volume);
  (b) Adjusting air conditioner (turn on air conditioning, set up temperature, turn on the ventilator for the specific zone);
  (c) Setting other commands (switching internal air circulation on/off).

(F) **Smoking** (n = 1, average duration 199 s): take a pack of cigarettes from the glove compartment, then take one cigarette out of the box, return the pack of cigarettes to the glove compartment, take a lighter and light a cigarette (or touch a cigarette with a lighter), put away the lighter, hold the cigarette with the dominant hand, for the end put it in the ashtray.

(G) **Drinking** (n = 1, average duration 167 s): with the dominant hand, the driver must retrieve the PVC bottle from the seat behind him or the stand next to him, open it, put the cap away, take a few sips of the drink, close the bottle, and return the PVC bottle to the same place where he took it (repeated 4 to 6 times, until the end of the section in the scenario is reached).

The 1120-meter-long route in ZUK Borongaj is divided into seven sections according to Fig. 1 and Table 1, of which only 13.5% is the length of the change of direction (four sharp turns to the left).

On the route in ZUK Borongaj shown in Fig. 1, there is interaction with public transport buses, teachers' and students' cars, and student pedestrians. The bus stops are located on sections 3 and 7, with pedestrian crossings just before the stop. At the four pedestrian crossings marked in Fig. 1, students simulated pedestrians in the presence of pedestrians, if necessary. The average speed of the vehicle in which the camera filmed the driver was 25 km/h, according to the data from the Endomondo platform.

The matrix of the driver's activities imposed by the scenario by sections and laps is shown in Table 2. The drivers drove seven laps according to the scenario shown in Table 2, including lap number zero, which represent the baseline driving without additional driver activities imposed by the scenario.

The drivers are told that they must not stop for no reason, that they must comply with traffic rules and regulations, drive at a speed at which they feel comfortable and safe, but in a way so that they can perform tasks.

Table 2 Matrix of driver's activities imposed by the scenario by sections and laps

Lap number	Label and full name of the section						
	Flat part west	Sharp left	Flat part south	Sharp left	Flat part east	Shart left	Flat part north
	1	2	3	4	5	6	7
0							
1	A		D1.1.	D1.1.	D1.1.	D1.1.	D1.1.
2	B3.1	B3.1	B3.1		E1	E1	E1
3	F	F	F	F	F	F	F
4			D1.2.	D1.2.	D1.2.	D1.2.	D1.2.
5	B3.2	B3.2	B3.2		E2	E2	E2
6			G	G	G	G	G

The empty fields in Table 2 imply the baseline driving without any driver's activities imposed by the scenario. Route sections with a change of direction (four sharp turns to the left at 90°) are not excluded from the analysis of the results using the driver's glance behavior method because:

(a) they take over 13.5% length of the route of one lap;
(b) baseline driving as well as all other activities of the driver imposed in the scenario from Table 2 also contain a change of direction due to driving through bends, except for activity (A) setting the navigation on the mobile phone.

Drivers were recorded by GOPRO HERO8 Black cameras, placed on the central part of the windshield, on the stand on the left and under the rearview mirror.

3 Results

For parameter gaze direction (number of glances in a zone) shown in Table 3, all zones with two or fewer glances during the duration of activity are excluded from the results. For the parameter percent of gaze on the road center, the information on the percentage share of gaze length in the main zone "Left side of the windshield" was used, which is also shown in Table 3.

Table 3 shows a high overlap of results for all three drivers from the age group of male respondents younger than 30. Although these results cannot be taken as final due to the insufficient number of respondents, they point to the justification of continuing the research on a larger number of samples of male drivers, six drivers in each of the three age groups:

Table 3 Percentage of gaze length shares and number of zones used depending on the driver's activities

		% of total gaze length[a]			Number of zones used		
		Driver 1	Driver 2	Driver 3	Driver 1	Driver 2	Driver 3
	Baseline driving	**77.2**	**75.1**	**70.8**	**8**	**9**	**10**
F	Smoking	76.3	65.4	64.4	9	9	6
G	Drinking	86.4	76.3	78.2	6	11	7
D1	Cognitive task through mobile phone conversation (video call)	88.9	73.6	70.4	5	6	5
A	Setting up navigation on a cell phone	17.3	23.1	32.9	2	2	3
E	Complex task	68.9	63.0	60.2	5	7	8
B3	Mobile phone conversation (video call)	74.3	57.4	51.3	3	5	3

[a]% of total gaze length in the main zone "Left side of the windshield"

(a) up to 30;
(b) from 31 to 60;
(c) over 60 years of age.

The following is a possible interpretation of the results shown in Table 3.

Driver's activities (F) Smoking, and (G) Drinking, are mostly manual distractions and the elements of visual distraction have in themselves the beginning and the end of the activity (reaching for a cigarette or a PVC bottle and disposing them at the end). According to the results shown in Table 3, they are very similar to baseline driving, especially in the segment of zones used. Therefore, such driver's activities cannot be reliably detected by using the driver's glance behavior method.

Driver's activity (D1) Cognitive task through mobile phone conversation (video call) is predominantly a cognitive distraction, and in accordance with the claims from literature [6] causes an increase in the percentage of gaze in the main zone "Left side of the windshield" (for Driver 1 only) with a moderate reduction in the number of zones used (for all drivers) in relation to the baseline driving.

Driver's activity (A) Setting up navigation on a cell phone is the most dangerous driver's activity, a mix of visual, manual and cognitive distraction, and in accordance with claims from literature [7] causes a significant decrease in the percentage of gaze in the main zone "Left side of the windshield" with a simultaneous increase in the percentage of gaze in the zone "Navigation (Central part of the windshield)" where the cell phone is fixed on the stand, as well as with the significant reduction in the number of zones used (only two above mentioned) in relation to the baseline driving.

Driver's activities (E) Complex task and (B3) Mobile phone conversation (video call) are very similar and they are probably predominantly visual distractions, and in accordance with the claims from literature [8] cause a decrease in the percentage of gaze in the main zone "Left side of the windshield" with a reduction in the number of zones used in relation to the baseline driving.

It is important to note that the percentage of gaze in the main zone "Left side of the windshield" shown in Table 3 is calculated as the mean value for the entire driver's activity; therefore, border values of this parameter inside of the value range can be different.

4 Conclusion

The authors of the paper are satisfied with the conducted research because they have mastered the methodology of measuring and processing the results, and human capacity for the larger-scale research has been ensured. In this paper, it is shown that using the driver's glance behavior method can be used to detect distractions that are predominantly visual such as driver's activities (E) Complex task and (B3) Mobile phone conversation (video call), in accordance with the claims and results from literature, emphasizing that the final results should be confirmed by the increased number of male subjects in the three age groups. The glance duration parameter, especially in the category of the border amount of the longest view for a particular driver's activity, will be covered with larger-scale future research on a larger number of respondents.

The preliminary results showed that the most dangerous driver's activity among all the researched activities in this preliminary study is setting up navigation on a cell phone placed on the stand on the left and under the rearview mirror, which results with a combination of visual, manual and cognitive distraction. The aforementioned driver's activity results in a combination of the two most dangerous possible consequences: a significant reduction in the percentage of gaze in the main zone "Left side of the windshield" with the significant reduction in the number of zones used (only two). Very similar activity during driving is writing a text message on the cell phone placed also on the stand under the rearview mirror; therefore, future research might be supplemented with the abovementioned activity.

Also, since the three respondents in this paper are male students who are amateur drivers, this paper does not take into account gender and occupation factors (amateurs or professional drivers with experience), so there is space for improvement in the research methodology in the larger-scale research planned for the near future.

References

1. Vasli B, Martin S, Triverdi MM (2016) On driver gaze estimation: explorations and fusion of geometric and data driven approaches. In: IEEE 19th international conference on intelligent transportation systems (ITSC). Rio de Janeiro, pp 655–660
2. Wang Y, Yuan G, Mi Z, Peng J, Ding X, Liang Z, Fu X (2019) Continuous driver's gaze zone estimation using RGB-D camera. Sensors 19(6)
3. Choi I-H, Hong SK, Kim Y-G (2016) Real-time categorization of driver's gaze zone using the deep learning techniques. In: 2016 International conference on big data and smart computing (BigComp). Hong Kong, pp 143–148
4. Vora S, Rangesh A, Trivedi MM (2018) Driver gaze zone estimation using convolutional neural networks: a general framework and ablative analysis. IEEE Trans Intell Veh 3(2):254–265
5. Caird et al (2018) Does talking on a cell phone, with a passenger, or dialing affect driving performance? An updated systematic review and meta-analysis of experimental studies. Hum Factors 60(1):101–133
6. Victor TW, Harbluk JL, Engström JA (2005) Sensitivity of eye-movement measures to in-vehicle task difficulty. Transp Res Part F: Traffic Psychol Behav 8(2 SPEC. ISS):167–190
7. Wang Y, Bao S, Du W, Zhirui Y, Sayer JR (2017) Examining drivers' eye glance patterns during distracted driving: insights from scanning randomness and glance transition matrix. J Saf Res 63:149–155
8. Sodhi M, Reimer B, Llamazares I (2002) Glance analysis of driver eye movements to evaluate distraction. Behav Res Methods Instrum Comput 34(4):529–538

Clothing Sizes of China and Japan and Contemporary Design Inspired by Elements of Folk Costume

Blaženka Brlobašić Šajatović and Dora Busak

Abstract The system of clothing sizes in China and Japan is described, and the folk costumes of China and Japan are presented. In traditional clothing, one wants to find influence and inspiration for creating contemporary design. This paper contains an experimental or art-technical part which consists of a fashion illustration of the entire collection, project drawings, construction and modeling of each model separately.

Keywords Anthropometric measurements · Clothing sizes · China · Japan · Traditional folk costumes · Fashion illustration

1 Introduction

Ergonomics is a scientific discipline (work science) whose task is to investigate the human body and behavior, and provides data on the suitability of objects with which man comes into contact. Thus, ergonomics studies the anatomical, physiological and other parameters of the human body. Anthropometric measures are applicable in various fields of scientific activity, both in ergonomics and in the textile industry. Anthropometry in ergonomics optimizes human interactions with equipment and the workplace. Because people are different in height and body development, a comprehensive study of the human body is needed. Dynamic anthropometry is especially important in determining the necessary measures for the construction of garments in which the body performs enhanced movements. This applies in particular to sports, work and protective clothing and to the uniforms of police and military units. Static anthropometry is performed according to the standards ISO 3635, ISO 8559 and ISO 9407, which laid the foundations for a unique definition of human body measures for the needs of the clothing and footwear industry, and thus the method of anthropometric measurements. Based on

B. B. Šajatović (✉) · D. Busak
University of Zagreb, Faculty of Textile Technology, Zagreb, Croatia
e-mail: blazenka.brlobasic@ttf.hr

D. Sumpor et al. (eds.), *Proceedings of the 8th International Ergonomics Conference*, Advances in Intelligent Systems and Computing 1313,
https://doi.org/10.1007/978-3-030-66937-9_5

anthropometric measurements and statistical processing of the obtained results of body measurements, a system of clothing sizes is defined. Different systems are used to mark clothing and footwear sizes worldwide, creating a range of concerns and difficulties for manufacturers. Over the years, clothing size systems have been created in different parts of the world, which vary widely in terms of the parameters on which they are based, the intervals of component size and, in particular, the way of designation of garment sizes [1].

2 Importance of Anthropometric Measurements

Interest in clothing size system will increase as the number of older users is expected to double by 2030. This poses a market challenge for the apparel industry as poor sizing is the main reason for the return and price reduction, resulting in significant losses. A 3D body scanner was used for anthropometric measurements in Japan (1992–1994). The measurement included about 19,000 Japanese men and 15,000 women aged 7–90 years. 178 measurements were obtained with a Voxelan laser three-dimensional body scanner and traditional procedures.

The main motivation for this sizing review was to understand the changes that have occurred in body size and shape in Japan. The average height of the Japanese has increased by more than 10 cm over the last 100 years. A research center for human engineering was established at the Department of Clothing Technology in Beijing, China in 1999. A comparison of the body types of women in college was conducted. A review of Chinese women's sizes was conducted in 2003, in collaboration with a local lingerie manufacturer. At the end of 2003, a Japanese Voxelan 3D scanner is installed. Between 1999, until 2000 a research project, "Studying Body Shapes for Comparison between Korea and China," was also conducted for Seoul University [2]. Anthropometric data of th e elderly have become an immediate need for ergonomic design of health care and living products even in a developing country like China. The first aim of this survey was to collect anthropometric data of the Chinese elderly (aged over 65) living in the Beijing area. 58 females (age range 65.0–80.7, mean 71.2, SD 4.1) and 50 males (age range 65.2–85.1, mean 71.5, SID 4.4) took part in the survey. A total of 47 anthropometric dimensions and three items of functional strength were measured. Mean values, standard deviations, coefficients of variation, and percentiles for each parameter were estimated. It was found that in most dimensions there were no significant differences between the age groups of 65–69 and 70–74 or between the age groups of 70–74 and 75+. Male and female elderly had no significant differences in the body dimensions around the hip area. Comparison between Chinese (Beijing) and Japanese elderly shows that Chinese (Beijing) elderly are larger in the dimensions of the body trunk, and Japanese elderly are larger in the dimensions of the head and extremities. The conclusions are based on a limited number of subjects in the Beijing area, and the in-depth reasons for the above findings remain a subject for further study. Relevance to industry the continuous growth of the number of aged people has created a big market of health care and living products for the elderly. Anthropometric data are essential to the ergonomic design of these products. However, available anthropometric data for aged people are quite limited. This study fills part of this gap by supplying anthropometric data of the Chinese elderly [3].

2.1 *Clothing Size System in China for Women*

There are two numbers in Chinese sizes, for example, size 88–90 is American size "S". This is because two different dimensions or parts of the body are measured, hip circumference and chest circumference. Residents of Western countries often have problems buying Chinese clothes, because Chinese standards are lower than European or American ones, so clothes are often too small for customers. There are different designations of clothing sizes in countries. The table shows a comparison of Clothing sizes for women between different countries with an emphasis on clothing sizes in China, Table 1. In China, they use two different systems. This is because they measure two different parts of the body. Japan uses the lettered system XS–XL. Japan also uses two different numbered systems in sizing, Tables 1 and 2 [1].

2.2 *The Folk Costume of China and Japan*

The folk costume of China was greatly influenced by the dynasties that ruled over the centuries, the ability of the population to make textile techniques, especially in silk, and the quality of the textile raw material itself. Ancient Chinese clothing or Hanfu refers to the historical styles of dress in China. Qipao or Cheongsam/Changshan traditional dress in the Manchu ethnic group. Qipao developed from

Table 1 Female clothing size conversion chart in China [4]

Female clothing size conversion chart								
Korean		44	55	66	77	88		
Japan	5	7	9	11	13	15	17	19
	SS, XS	S	M	L	LL, XL	XXL, 3L	4L	5L
		36	38	40	42			
International		XS	S	M	L	XL	XXL	
US		0–2	4–6	8–10	12–14	16–18		
UK		6–8	8–10	12–14	16–18	20–26		
France		32	34–36	38–40	42–44	46–48		
Italy		36	38–40	42–44	46–48	50–52		
China			160–165	165–170	167–172	168–173	170–176	
			84–86	88–90	92–96	98–102		

Table 2 Marking of clothing size of women's clothing [4]

Women's clothing								
Generic		XS	S	M	LL, XL	XL	XXL	
Korea		44	55	66	77	88		
	5	7	9	11	13	15	17	19
Japan	SS, XL	S	M	L	LL, XL	XXL, 3L	4L	5L
		36	38	40	42			
USA		0–2	04–06	08–10	12–14	16–18		
UK		32	34–36	38–40	42–44	46–48		
Europe		36	38–40	42–44	46–48	50–52		

female changpao (long dress) during the reign of the Qing dynasty (Fig. 1) Qipao is a tight dress that "hugs" a woman's body, and contains characteristic features of Manchu origin [5]. Class divisions and boundaries in Japan were strict, and one easy way to support divisions was through the use of ancillary laws that restricted the use of certain fabrics and colors or styles of clothing to those in different classes. The clothing of the elite was strictly controlled and had great significance, symbolizing the place where each person ranked in society relative to others. Clothing highlighted a person's wealth or position in government/society. Folk costumes in Japan as well as in China are divided into more famous clothing items and those less known, of which the most famous is the Kimono (Fig. 2). The least formal

Fig. 1 Women's Qipao dress [7]

Fig. 2 Kimono [8]

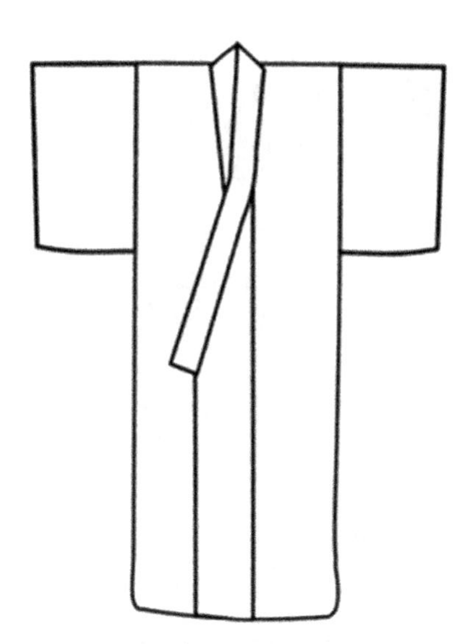

kimono is made from other types of fibers, including warm wool, cold cotton or linen, but also from fibers obtained by new technologies, such as man-made fibers used as cheaper alternatives to expensive silks used in the past. For women, the design of the kimono and the color must be chosen with great care and attention to her marital status and age. As in previous days, the festive kimono is often brightly colored with colorful patterns and a design that contains great meaning. The construction with the longest sleeves reaches the lower edge of the kimono, close to the length of the floor, and is worn by brides for weddings, and shorter sleeve lengths are worn by guests [6].

3 Display of Fashion Illustration, Construction and Modeling

Collections inspired by elements of the folk costumes of China and Japan are shown. Model inspired by traditional Qipao dresses with contemporary interventions, (Figs. 3, 4 and 5) [8]. The pictures show the dresses and their construction and modeling.

Fig. 3 Model 1 [8]

Fig. 4 Model 2 [8]

Fig. 5 Model 3 [8]

4 Conclusion

The importance of anthropometric measurements finds its purpose in the textile industry as well. Based on the statistical processing of body measures, a system of clothing sizes of individual countries is obtained. In China, they use two different systems. This is because they measure two different parts of the body. Japan uses the lettered system XS–XL. Japan also uses two different numbered systems in sizing. In the experimental part, the inspiration comes from the need to fit the segments of folk costumes of China and Japan into contemporary design, and from the need to play with cuts and patterns.

References

1. Ujević D, Brlobašić Šajatović B, Bin Y, Feichao Z (2020) Comparasion of croatian and Chinese clothing size system with reference to characteristic body proportions. In: Petrak S, Zdraveva E, Mijović B (eds) 13 th International scientific—professional symposium, textile science and economy. University of Zagreb Faculty of Textile Technology Zagreb, pp 98–103
2. Ujević D (2010) Theoretical aspects and application of croatian anthropometric system. University of Zagreb Faculty of Textile Technology, Zagreb
3. Hu HT, Zhizhong L, Yan JB, Wang XF, Duan JY, Lei Z (2007) Anthropometric measurement of the Chinese elderly living in the Beijing area. Int J Ind Ergon 4:303–311
4. https://womendressideas2019.blogspot.com/2019/03/women-clothing-us-size-12-in-china.html . Accessed 1 Nov 2020
5. Brown J (2006) China, Japan, and Korea: culture and customs. Creatspace Independent Publishing, South Carolina
6. Yang S (2004) Chinese clothing: costumes, adomments and culture (Arts of China). Long River Press, San Francisco
7. https://YueLian-Womens-Chinese-Evening-Cheongsam/dp/B00PIKW0CS. Accessed 1 Nov 2020
8. Busak D (2020) Utjecaj tradicionalne narodne nošnje Kine i Japana na suvremeni dizajn. University of Zagreb, Faculty of Textile Technology, Zagreb

Development of Ergonomic Sportswear Based on Thermal Body Mapping

Ivana Salopek Čubrić, G. Čubrić, and V. M. Potočić Matković

Abstract The aim of the research is to get insight into the changes of body temperature of an athlete in order to perform detailed body mapping. Developed body maps are to be further used for design of ergonomic sportswear for a futsal player. In the experimental part in this paper is used thermal camera FLIR E6. The measurement is conducted with participation of active futsal player, before and after a number of training sessions. The measurement is carried in standard conditions for a futsal training, i.e. indoor environment. The measured values of temperature are observed before and after intensive exercise for the zones of upper and lower body and used to develop body maps. The outcomes of the research are to be used for the development of personalized ergonomic futsal sportswear.

Keywords Sportswear · Futsal · Thermography · Thermal comfort · Ergonomy

1 Introduction

Sports activities are a trend, more and more people are engaged in physical activity. Futsal is a sport of the modern age. The equipment at first glance does not require too much: T-shirts, shorts, socks, sneakers. Futsal is a very physically intense sport, evidence from the literature shows that the physical demands of futsal are important consideration. Castagna et al. [1] examined the physiological responses and activity

I. Salopek Čubrić · V. M. Potočić Matković
Department of Textile Design and Management, University of Zagreb Faculty of Textile Technology, Zagreb, Croatia
e-mail: ivana.salopek@ttf.unizg.hr

V. M. Potočić Matković
e-mail: marija.potocic@ttf.unizg.hr

G. Čubrić (✉)
Department of Clothing Technology, University of Zagreb Faculty of Textile Technology, Zagreb, Croatia
e-mail: goran.cubric@ttf.unizg.hr

D. Sumpor et al. (eds.), *Proceedings of the 8th International Ergonomics Conference*, Advances in Intelligent Systems and Computing 1313,
https://doi.org/10.1007/978-3-030-66937-9_6

pattern for futsal simulated game-play in professional players, and found that futsal played at a professional level is a highly demanding physically intense exercise which stretches the aerobic and anaerobic capabilities of players. Moore's et al. [2] research related to the futsal literature include diverse subjects as coaching, physiological, psychological, technical and tactical elements of the sport, development of futsal and the prevalence of injuries. A review of the literature shows that research on the comfort of players' clothes is missing.

For the last decades, scientists have studied different testing methods for measurement of comfort of sports clothes. The comfort of a garment can be divided into three groups: psychological, tactile (sensory) and thermal comfort. The methods of measuring thermal comfort can be divided into two main groups:

(a). Method for measurement using thermal manikins (i.e. objective method).
(b). Wear trials using human participants (persons in a real environment, measuring with a thermometer or infrared thermography) [3].

Infrared thermography is gaining popularity among the researchers in various fields. In this investigation, thermography is used to observe the changes in skin temperature of futsal players wearing athletic gear. For the measurement is used thermal camera FLIR E5 [4].

During sports activity, the core temperature of the body rises, depending on the energy obtained by biochemical reactions. Limited amounts of energy are converted into mechanical energy, while the rest is converted into thermal energy, i.e. most of the energy is released in the muscles as heat [5, 6]. The lowest energy required, the basal metabolic rate, depends on gender, age and body temperature [7]. The human body can be easily thermoregulated. Physiological thermoregulation is triggered by temperature signals from the body and skin, and behavioural thermoregulation is associated with dressing and removing clothes, which is closely related to the thermal comfort of clothes [5, 6]. Heat is usually transferred by convection (by the flow of air) from the skin to the environment, because the environment is usually colder. Between the body surface and the environment, there will be heat exchange also by radiation and conduction. The body furthermore possesses the possibility of heat loss by evaporation. Moisture from the skin evaporates causing a large amount of heat to be released. Heat loss also occurs in the lungs because the inhaled air is usually colder and drier comparative to the interior of the lung surface [5].

The transfer of air, heat and water vapor through textiles are probably the most important factors in clothing comfort. There is no extensive research related to the comfort of futsal players' equipment, but numerous studies have reported on various aspects of comfort-related investigations of knitted textile products. Thermography was used to observe influences of different types of yarns to the total drying time of knitted fabrics. The results indicated significant differences in total time of drying between fabrics. The influence of the elastane yarn to the total drying time directly affect human comfort is proven [8]. The structure of yarn (of the same fibre and the same fineness) has great impact on the comfort related properties, i.e. air permeability, thermal conductivity, percentage water vapor permeability and water absorbency [9]. Thermal properties of knitted fabrics made from cotton and regenerated bamboo cellulosic fibres were examined. It was found that the thermal conductivity of knitted fabrics generally reduces as the proportion of bamboo fibre increases. For the same fibre blend proportion, the thermal conductivity was lower for fabrics made from finer yarns. The thermal conductivity and thermal resistance values of interlock fabric was the maximum followed by the rib and plain fabrics. The water vapour permeability and air permeability of knitted fabrics increase as the proportion of bamboo fibre increases [10]. Nemeckova et al. [11] used the method of measuring moisture transport on knitwear using thermographic and microthermographic systems. Three types of knitwear of the same raw material composition and fineness, but different construction were examined, among which interlock has been shown to give the best results in terms of liquid absorption and transfer. Mass per unit area and thickness of knitted fabric influence the water vapor resistance. The results did not indicate the influence of knitted fabric density. Power-net and voile structure allow maximal transfer of water vapor to the environment; and simplex/locknit decrease water vapor transfer. A very good correlation of mass per unit area of knitted fabrics, and thermal resistance of knitted fabric occurred [12, 13]. Thermography was used to observe that drying time of single jersey cotton fabric without elastane may be up to 70 min shorter than drying time of similar fabrics with elastane used for the same purpose, which directly affects human comfort [14].

In the field of sport, thermographic studies often report their quantitative findings by body mapping of skin temperature. This technique enables analysis of several thermograms and provides population-average whole-body maps of skin temperature. Averaging pixels over the entire body gives a true estimation of the mean skin temperature for the group of measurements [15].

The aim of the research presented in this paper is to get insight into the changes of body temperature of futsal player in order to develop thermal body maps and use them for further design of highly-ergonomic outfit.

2 Experimental

In the experimental part of the study participated active futsal player in good health condition. The measurements of player's body is conducted during the period of two months, before and after trainings of his team. There were three types of typical trainings:

a. elements of futsal game only
b. strength and endurance and
c. combined training of stretching and elements of futsal game.

Each training lasted for 60 min. During the training, the player was dressed in official futsal club's clothing produced of 100% polyester that consisted of short sleeved shirt and shorts. For the measurement of player's body temperatures, the

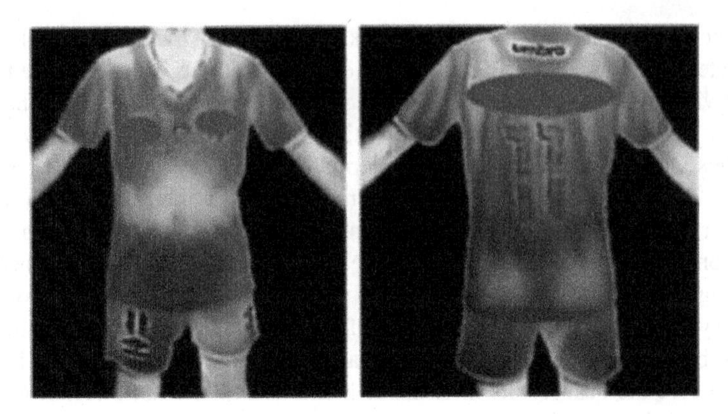

Fig. 1 Example of thermograms (anterior and posterior)

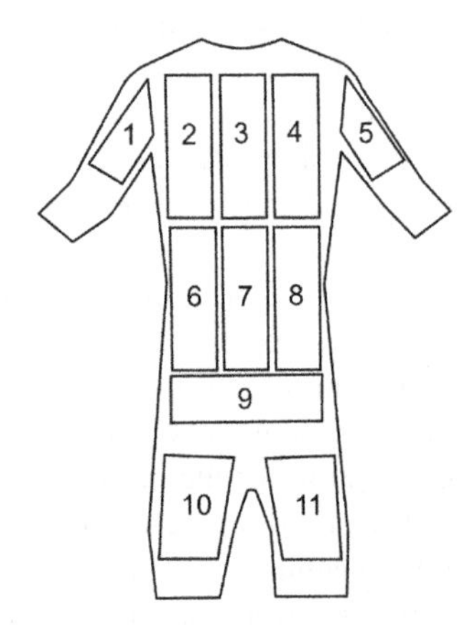

Fig. 2 Overview of body zones (same layout for anterior/posterior)

thermal camera FLIR E6 is used (example of thermograms is given in Fig. 1). The measurement is carried out in indoor environment with temperature t = 22 ± 0.5 °C and relative humidity 55 ± 5%. The values of temperatures are further processed in Flir Tools professional software for thermal analysis. In the focus of interest were 22 body zones (11 anterior and 11 posterior) as shown in Fig. 2. The average values of temperatures are based on 5 measurements for each type of training. The concept of the presented research is conducted with approval of the Committee of Ethics of the University of Zagreb, Faculty of Textile Technology.

3 Results and Discussion

The results of measured temperatures are given in Fig. 3. As can be seen from Fig. 3, after the trainings that focus at elements of game only, the measured temperatures slightly increase for the majority of anterior zones. The significant

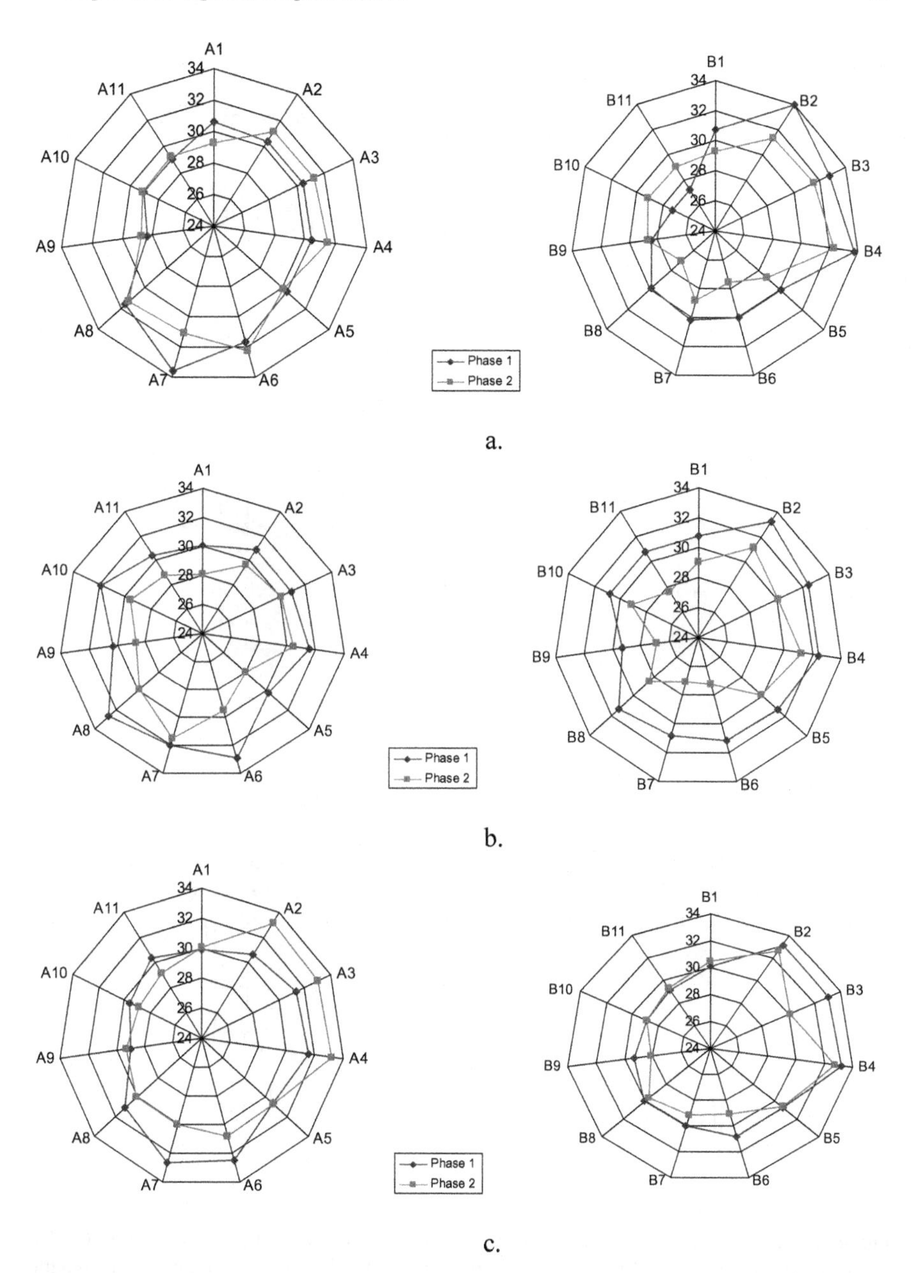

Fig. 3 The results of measured temperatures before (Phase 1) and after (Phase 2) conducted types of trainings: **a** training of elements of futsal game only, **b** training of strength and endurance and **c** combined training of stretching and elements of futsal game

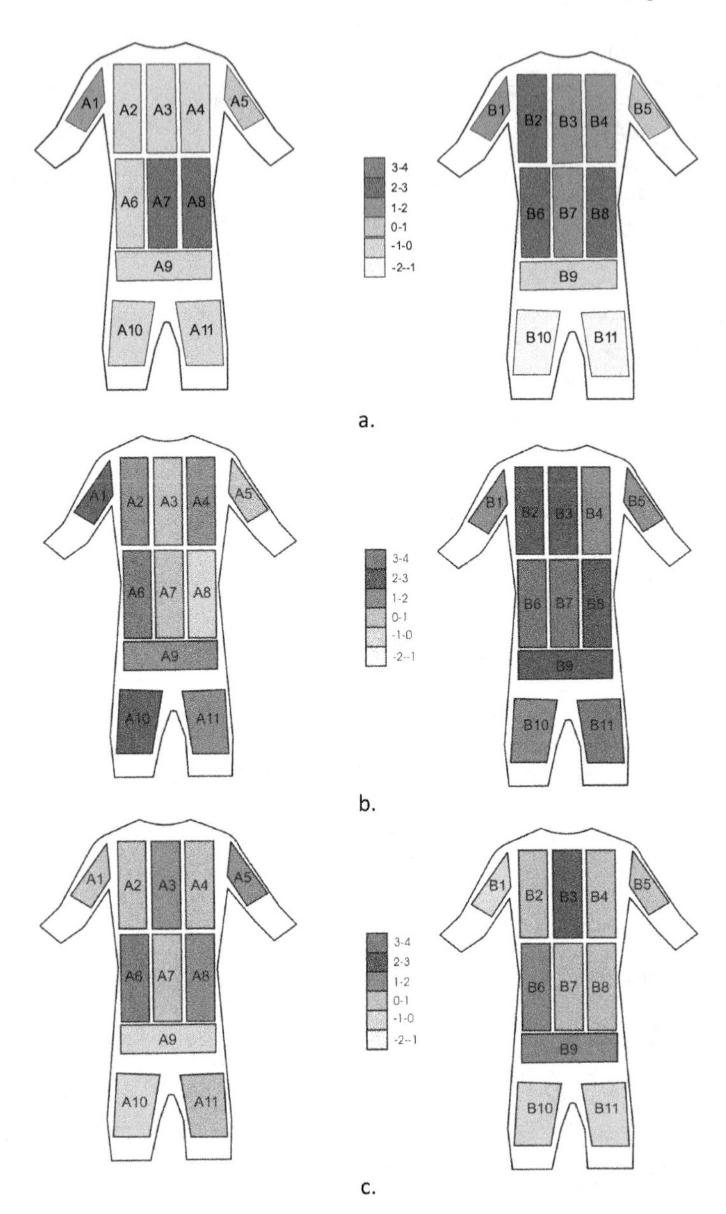

Fig. 4 Body maps showing the differences in body temperatures after trainings (in °C) for different types of futsal training: **a** training of elements of futsal game only, **b** training of strength and endurance and **c** combined training of stretching and elements of futsal game

decrease of temperature is noted for the zone 7, i.e. the zone of medial abdomen. After the training of strength and endurance, the temperature decreases for all observed anterior zones. This type of training is very complex and demanding, what activated intensive body sweating. Due to the fact that 100% PES fabric does not have absorbing capability, the sweat is accumulated on the skin surface and this affected the decrease of measured temperature. Finally, after the combined training of stretching and elements of futsal game, the temperature increases for the zones A1–A5, i.e. upper chest and upper arm zones. The results implicate that the body warmed up due to the training activity, but the mechanism of body sweating did not activate yet. For all three types of training, the temperatures of posterior zones decrease after completion of training. The exception is the training dedicated to the elements of game only, where the temperatures slightly increase for the zones of lower back and upper tight.

The body maps showing the differences in body temperatures after trainings for three different types of futsal training are shown in Fig. 4.

As can be seen form presented body maps, there are significant differences between body maps developed after different types of trainings. When compared, it can be concluded that there are no uniform patterns regarding the observed zones. In principle, for the anterior zones, there is an increase of temperature in the zones of abdomen (zones A6–A8). It is also interesting to observe the differences in the body maps for the zones of upper tight. As can be seen, there is a significant decrease of temperature after the training of the elements of futsal game only. As opposite, the significant increase of the temperature after the strength and endurance training.

4 Conclusion

The results presented in this paper focus on the measurement of surface temperature of participating futsal player after a three types of trainings. The outcomes of the research outlined the importance of adjustment of the construction and fit of the clothing to the type of training or activity being performed in order to fulfil ergonomic requirements. The results of thermographic measurements gave guidelines how to proceed with design of clothing for trainings of three different levels of intensity. The presented research and its results further indicate the importance of developing personalized clothing for top athletes, which in today's age, when better sport results are always expected, is imposed as an imperative.

Acknowledgements This research is supported by University of Zagreb, within research grants TP12/20 "Design of functional sportswear with implication of mathematical models and algoritms".

References

1. Castagna C, D'Ottavio S, Vera JG, Álvarez J, Carlos B (2009) Match demands of professional futsal: a case study. J Sci Med Sport 12(4):490–494
2. Moore R, Bullough S, Goldsmith S, Edmondson L (2014) A systematic review of Futsal literature. Am J Sports Sci Med 2(3):108–116
3. Potočić Matković VM, Salopek Čubrić I, Čubrić G (2019) Performance of diving suits from the aspect of thermal comfort. In: Novak I, Schwarz I, Špelić I, Zdraveva E (eds) Proceedings of the 12th scientific-professional symposium textile science and economy. University of Zagreb Faculty of Textile Technology, Zagreb, pp 69–74
4. Flir Homepage, http://www.flir.com/instruments/ex-series/. Accessed 14 Sept 2020
5. Havenith G (1999) Heat balance when wearing protective clothing. Ann Occup Hyg 43 (5):289–296
6. Zora S, Balci GA, Colakoglu M, Basaran T (2017) Associations between thermal and physiological responses of human body during exercise. Sports 5(4):97
7. Parsons K (2003) Human thermal environments. Taylor and Francis, London
8. Salopek Čubrić I, Čubrić G, Potočić Matković VM (2016) Use of thermography to analize the influence of yarn properties on fabric drying. In: Guxho G (eds) Book of proceedings of the 7th international textile conference, Tirana 2016. University of Tirana, Tirana, pp 367–371
9. Das A, Ishtiaque SM (2004) Comfort characteristics of fabrics containing twist-less and hollow fibrous assemblies in weft. J Text Apparel Technol Manage 3(4):1–5
10. Majumdar A, Mukhopadhyay S, Yadav R (2010) Thermal properties of knitted fabrics made from cotton and regenerated bamboo cellulosic fibers. Int J Therm Sci 49(10):2042–2048
11. Nemcokova R, Glombikova V, Komarkova P (2015) Study on liquid moisture transport of knitted fabrics by means of Mmt, thermography and microtomography systems. AUTEX Res J 15(4):233–242
12. Salopek Čubrić I, Potočić Matković VM, Skenderi Z, Tarbuk A (2017) Impact of substrate on water vapor resistance of naturally weathered coated fabrics. Text Res J 87(13):1541–1553
13. Potočić Matković VM, Salopek Čubrić I, Skenderi Z (2014) Thermal resistance of polyurethane-coated knitted fabrics before and after weathering. Text Res J 84(19): 2015–2025
14. Salopek Čubrić I, Čubrić G, Potočić Matković VM (2016) Thermography assisted analysis of textile materials as a predictor of human comfort. In: Sušić A, Jurčević Lulić T, Salopek Čubrić I, Sumpor D, Čubrić G (eds) Proceedings of 6th international ergonomics conference, Ergonomics 2016. Croatian Ergonomics Society, Zagreb, pp 295–300
15. Fournet D, Redortier B, Havenith G (2012) A method for whole-body human skin temperature mapping in humans. Thermol Int 22(4):157–159

Achilles Tendon Load at Squatting Working Position

Maja Šimunić, Daria Ćurko, Tanja Jurčević Lulić, and Jasna Leder Horina

Abstract Workers often take squatting positions during work, some stay in this position for a short time, and some for longer, and often have an additional load in their hands (tools, instruments, loads). The aim of the study was to determine the load that occurs in the Achilles tendon while squatting. The Achilles tendon is one of the largest and strongest tendons, but too much stress can cause pain, swelling, tendinopathy, tendon damage or even rupture. A mathematical model based on the statics of rigid bodies was set up to determine the force in the Achilles tendon. Experimental part of the study was conducted using the Elite system and the Zebris platform. The loads of the Achilles tendon when squatting on the full foot and on the toes were determined. The determined stress values in the tendon are smaller than the values at which tendon rupture occurs, but still large values that can cause pain and pathological processes.

Keywords Squatting working position · Achilles tendon · Mathematical model · Stress

1 Introduction

The Achilles tendon connects the heel to the two muscles of the back of the lower leg and is the strongest tendon in the human body. During work, workers often take squatting positions, having an additional load in their hands (tools, instruments, different type of load). Due to overexertion of the Achilles tendon during squatting, pain, swelling, tendinopathy and tendon damage or even rupture can occur. There are different variations of tendinopathy, that is, tendon injuries due to wear and tear.

M. Šimunić · D. Ćurko (✉) · T. J. Lulić
Faculty of Mechanical Engineering and Naval Architecture, University of Zagreb, Zagreb, Croatia
e-mail: daria.curko@fsb.hr

J. L. Horina
Department of Mechanical Engineering, University North, Varaždin, Croatia

D. Sumpor et al. (eds.), *Proceedings of the 8th International Ergonomics Conference*, Advances in Intelligent Systems and Computing 1313,
https://doi.org/10.1007/978-3-030-66937-9_7

One of the most common tendon injuries in athletes is the one that occurs due to overload, side effects of medications, an accident, or due to wearing inadequate footwear. This type of injury results in inflammation, damage and weakening of the tendon which can eventually lead to tendon rupture. The places where inflammation most often occurs are the connection of the tendon to the heel bone and the very middle of the tendon.

Measurements have shown that there are differences in the elastic properties and stiffness of the Achilles tendon between individuals [1]. During activity, the behavior of the Achilles tendon is elastic and viscous, and the structure is viscoelastic because it consists of collagen fibers of solid matter and water. The mechanical properties of the Achilles tendon depend on the orientation and diameter of the collagen fibers and vary depending on the requirements that the tendon must meet. At rest, the fibers are arranged in a parallel manner, however, when they are stretched, they are straightened to the limit of elasticity and then they return to their initial state. Their main role is to transfer the axial load.

According to data from the literature [1–4], the modulus of elasticity of the Achilles tendon is 65 MPa, and the tensile strength is 24–61 MPa. A tendon with a cross section of 6 cm^2 can withstand a force of 40–80 kN. Although tendons can withstand heavy loads, due to poor blood supply they are very sensitive to inflammation and the tissue may become necrotic [1]. The geometric characteristics of the Achilles tendon can be precisely determined by ultrasound imaging. The narrower part of the tendon is about 4 cm wide and about 15 cm long, but even these data vary from person to person.

The advantage of a tendon is that it can adapt to a mechanical load. The modulus of elasticity and strength of the tendon can be increased by exercising, compared to the values found in the literature, which state the modulus of elasticity in the range of 500–1850 MPa [2], while immobilization and rest can lead to a decrease in these values.

Because the Achilles tendon is exposed to great forces during daily activities, it is prone to pathological changes called tendinosis which can be acute and chronic. The first symptom that occurs most often is pain.

The aim of the research is to use a simple method to estimate the load on the Achilles tendon that occurs while squatting, as one of the possible positions that a person takes during work.

2 The Method of Calculating the Load in the Achilles Tendon

A model that will be used to calculate the forces in the Achilles tendon at rest in the squat position consisting of: the Achilles tendon connecting the heel and the gastrocnemius muscle, the gastrocnemius muscle attached to the Achilles tendon and the tibia, the tibia extending from the ankle to the knee, and a foot represented by a triangle whose vertices are marked by points P (heel), G (ankle) and N (toe) [1]. Figure 1 shows a two-segment foot-shin model representing the connection between the Achilles tendon, gastrocnemius muscle, heel and tibia, while Table 1 shows the image nomenclature [1].

To determine the force in the Achilles tendon, it is necessary to know the anthropometers and external load that act on the subject. The experimental part of the research was conducted using the *Elite system* and the *Zebris* platform. The *Elite system* (*BTS Bioengineering*, Milan) enables monitoring and determination of the spatial coordinates of markers, which represent marked points on the body. The system consists of 8 cameras operating at a frequency of 100 Hz. The cameras have built-in flashes that emit infrared (IR) rays and are synchronized with the cameras. In the room there are also two normal cameras recording measurements from two different angles. IR cameras track markers and receive reflected infrared rays, converting them into electrons and into a digital record that is sent back to the computer's processor [5]. The *Zebris* platform is suitable for analyzing the reaction force of the ground at static body positions, but also while walking and running. This technology makes it possible to measure the pressure over the entire surface of the foot and determine the center of pressure (CoP) in which the resultant force of the ground is located.

Before recording and using this system, it is necessary to calibrate the camera system and the platform itself. It is also necessary to define the working volume, i.e. the space in which the measurement is performed. At the end of the system calibration, the anthropometers of the subject are measured and the data is then entered in the system. The subject is a 23-year-old woman (body mass 62 kg) with no history of foot-related disease. Programs used during imaging are *Elite Clinic* and

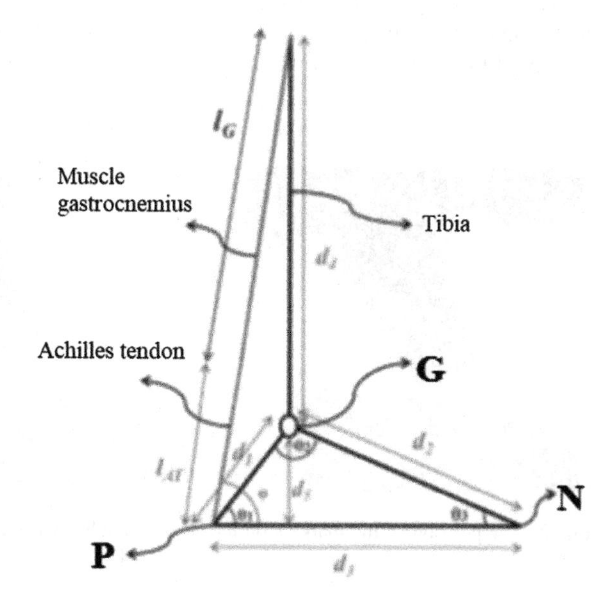

Fig. 1 A biomechanical model used to calculate the forces in the Achilles tendon [1]

Table 1 List of image nomenclature

Symbol	
d_1 (mm)	The distance between the heel (P) and the ankle (G)
d_2 (mm)	The distance between the ankle (G) and the toe (N)
d_3(mm)	The distance between the heel (P) and the toe (N)
d_4(mm)	The tibia bone length
d_5(mm)	The ankle distance (G) from the floor
l_G(mm)	Length of the gastrocnemius muscle
l_{AT}(mm)	The Achilles tendon length

Source Taken from Chatzistefanou [1]

SMARTanalyzer. Before shooting, it is necessary to place markers in certain positions on the foot. Recording is performed using 6 markers. One marker is placed on the inside and one on the outside of the ankle (*gležanj/vani, gležanj/unutra*). Two markers are placed so that the center of gravity of the foot is on their connecting line (*težište/vani, težište/unutra*). Another marker is on the connection of the Achilles tendon and the heel bone (*Ahilova/dolje*) and the last marker is located at the junction of the Achilles tendon and the gastrocnemius muscle (*Ahilova/gore*). Due to the assumed symmetry of both legs when squatting, it is sufficient to place the markers on only one leg. Figure 2 shows the layout of the markers placed on the right foot. The markers on the back determine the direction of the Achilles tendon, while the markers on the inside and outside of the ankle determine the center of rotation of the ankle itself. The remaining two markers indicate the position of the center of gravity of the foot which is in the middle of the connecting line of these two markers.

The recordings were performed while squatting on a full foot, followed by squatting on toes, all the while observing only the right foot (Fig. 3). The center of gravity of the foot is at 44.14% of the length of the foot looking from the heel to the

Fig. 2 The position of the markers on the subject's right foot; full foot on the floor

Fig. 3 The position of the markers on the subject's right foot; squatting with full foot on the floor versus squatting on toes

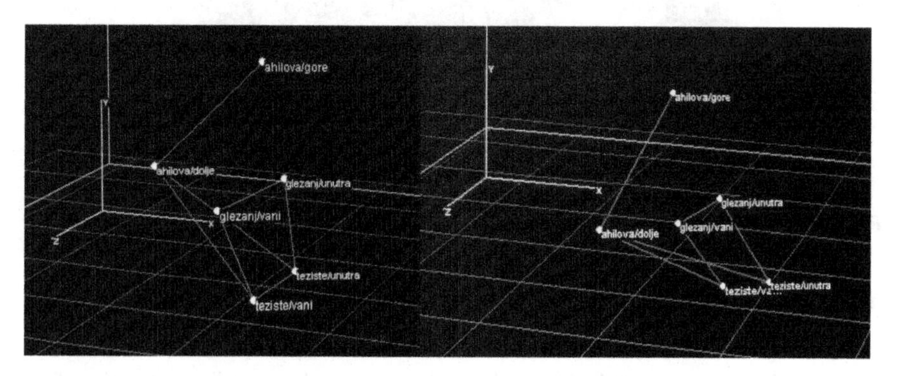

Fig. 4 Wire models for squatting on toes (left) and squatting on full foot (right)

toes [6]. The Y axis is perpendicular to the ground and the positive direction is from the foot to the knee. The X axis is the longitudinal axis whose positive direction is from the heel to the toes, and the Z axis is the transverse axis. The recording begins with the subject standing on the platform and taking a squatting position on full feet. The second recording is done in a squat on the toes. The software provides marker coordinates and wire models for both squats (Fig. 4).

During squatting on the Zebris platform, pressure distributions over the surface of the foot were determined. The pressure on the surface of the foot is shown in Fig. 5; the left part of Figure shows the pressure when squatting on a full foot, and the right the pressure of the foot when squatting on the toes. The concentration of forces on the front and back of the foot is also visible, and the center of pressure is indicated.

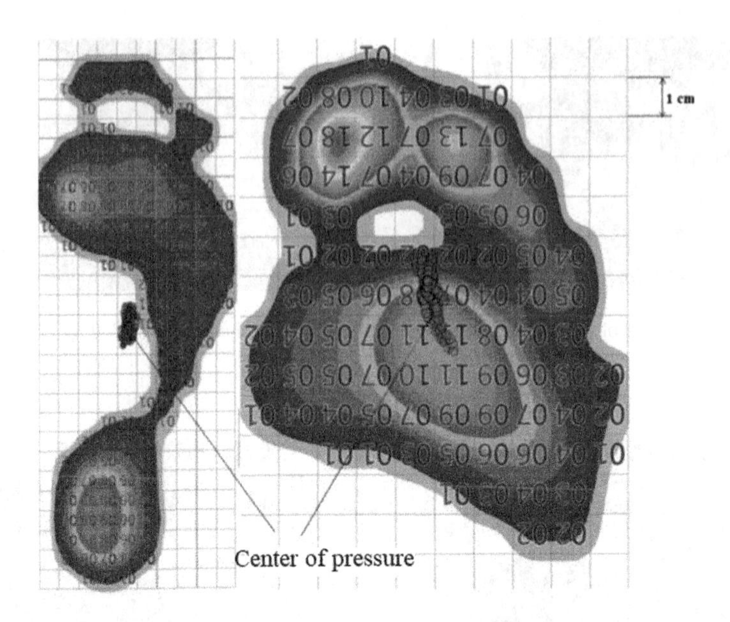

Fig. 5 Centers of pressure for squatting on full foot (left) and squatting on toes (right)

3 The Results

By recording squats on full foot and squats on toes, the coordinates of the characteristic positions of the markers, the magnitude and the point of application of the ground reaction force were determined. When squatting on a full foot, the center of pressure is at 135 mm from the heel. When squatting on toes, the center of pressure of the foot is 65 mm away from the toes along the Y axis.

To determine the force in the Achilles tendon, it is necessary to know the mass of the foot. The mass of the foot is $m_s = 0.9256$ kg. This was determined by immersing the foot in a beaker of water to determine its volume and multiplying it by the density of 1.095 kg/dm^3 [6].

The force in the Achilles tendon was determined on a two-segment model by setting the moment equation $\Sigma M_G = 0$ around the ankle joint, point G (see Figs. 6 and 7). When squatting on a full foot, the force in the Achilles tendon $F_{Ah} = 925.19$ N was calculated, while the force of $F_{Ah} = 166.96$ N was calculated when squatting on toes.

The force in the Achilles tendon, when squatting on a full foot, is about six times greater than the force during squatting on the toes. Assuming a tendon cross-section of 0.5 cm^2, the stress in the tendon during the full foot squat is 18.5 MPa, while it is 3.3 MPa when squatting on the toes.

It can be concluded that greater damage to the Achilles tendon could occur by squatting on a full foot. Greater damage can also lead to more serious tendon injuries.

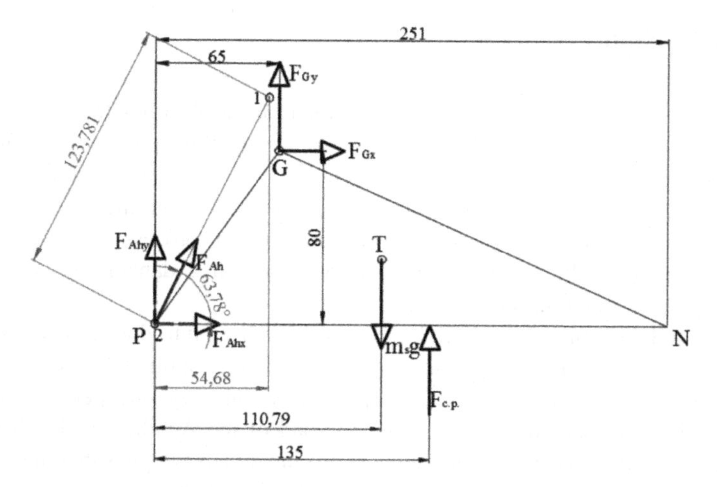

Fig. 6 Sketch for the momentum equation for full foot squat

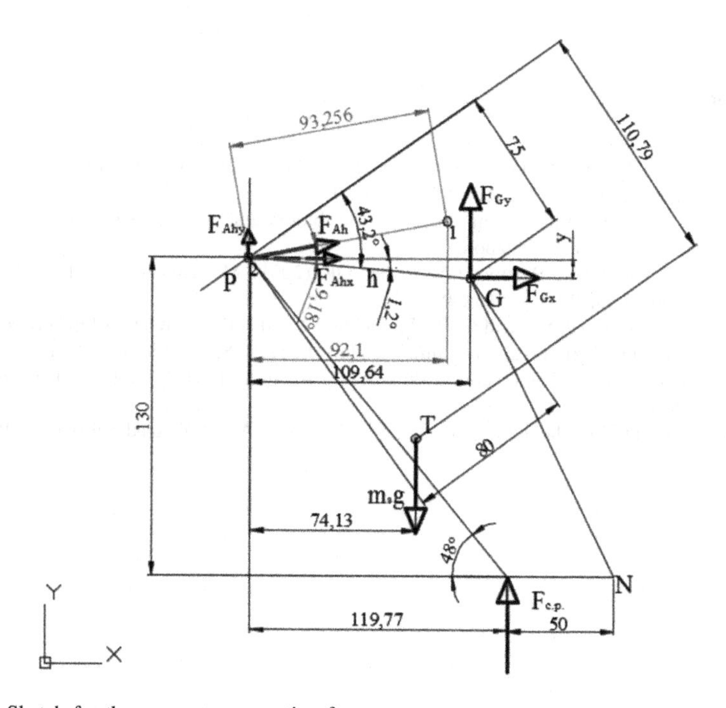

Fig. 7 Sketch for the momentum equation for squat on toes

4 Conclusion

From the results, we can conclude that the load on the Achilles tendon is greater when squatting on full feet than on the toes, so squatting positions on full feet should be avoided. The determined stresses in the tendon are less than the values at which tendon rupture occurs, however, it is still about the magnitude of stresses that can cause discomfort, pain and trigger pathological processes. Although the Achilles tendon is the strongest tendon, there is still a danger of its damage, especially when squatting on full feet.

In this study, simplifications and limitations have been introduced in determining the forces in the Achilles tendon. Rigid body statics were applied, footwear was not considered, and the measurement was performed on only one person. We can conclude that the method of calculating the load of the Achilles tendon has both advantages and disadvantages, but the most prominent advantage is that the loads of the Achilles tendon are easily estimated, which could help prevent major tendon injuries and thus reduce treatment costs.

References

1. Chatzistefanou N (2017) Mathematical modelling and simulation of the foot with specific application to the Achilles tendon. PhD thesis, The University of Warwick
2. Swee Hin T (2004) Engineering materials for biomedical applications. World Scientific Publishing Co. Pte. Ltd., Singapore
3. Wren TAL, Yerby SA, Beaupre GS, Carter DR (2001) Mechanical properties of human achilles tendon. Clin Biomech 16:245–251
4. Gomes J, Neto T, Vaz J, Schoenfeld BJ, Freitas SR (2020) Is there a relationship between back squat depth, ankle flexibility, and Achilles tendon stiffness? Sports Biomech 19:1–14
5. Šimunić M (2020) Achilles tendon loading during squating. Bachelor's thesis, University of Zagreb, FAMENA
6. Dempster WT (1955) The anthropometry of body action. Ann NY Acad Sci 63(4):559–585

Ergonomic Assessment with a Convolutional Neural Network. A Case Study with OWAS

Helios De Rosario, Enrique Medina-Ripoll, José Francisco Pedrero-Sánchez, Mercedes Sanchís-Almenara, Albert Valls-Molist, and Pedro Pablo Miralles-Garcera

Abstract The ergonomic assessment of workplaces is key to optimize work stations and tasks, and in some legal frameworks it is mandatory for companies to perform such evaluations. Currently, most of the evaluations are carried out by means of observational methods like REBA, OWAS, RULA, etc. The usual procedure to apply those methods consists of recording the worker in a normal work day, and then analyze postures, joint angles and movements by visual inspection. That is a tedious and slow process, which is very dependent on the evaluator's experience. Artificial vision can help to analyze and objectify the video in a few minutes. The objective of this work is the development of an ergonomic assessment method based on artificial vision and a convolutional neural networks, in order to reduce the time of analysis of the videos used in ergonomic evaluations. For this purpose, the neural network used (Simple Pose network) and its application to the analysis of postures is presented, and then a use case is presented, evaluating a workplace with the OWAS methodology.

Keywords Ergonomic assessment · Workplace · Artificial vision · Neural networks

1 Introduction

According to the last study of the European Occupational Safety and Health Agency (EU-OSHA), approximately three out of every five workers in EU-28 have symptoms related to musculoskeletal disorders (MSKD). The most frequent disorders of

H. De Rosario (✉) · E. Medina-Ripoll · J. F. Pedrero-Sánchez · M. Sanchís-Almenara
Instituto de Biomecánica de Valencia, Universitat Politècnica de València, València, Spain
e-mail: helios.derosario@ibv.org

A. Valls-Molist · P. P. Miralles-Garcera
Quirón Prevención S.L.U, Madrid, Spain

H. De Rosario
CIBER de Bioingeniería, Biomateriales y Nanomedicina, Zaragoza, Spain

D. Sumpor et al. (eds.), *Proceedings of the 8th International Ergonomics Conference*, Advances in Intelligent Systems and Computing 1313,
https://doi.org/10.1007/978-3-030-66937-9_8

that type among workers are pain in the back, neck and upper limbs, which is reported as the most serious problem by 60% of workers suffering some work-related health issue, and often develops as chronic pain [1]. The same study indicates that the main causes for MSKD are: manipulation of loads, specially when the task requires adopting flexed and twisted postures, repetitive or sudden movements, strained and static postures, vibrations, inadequate lighting or temperatures, fast work cycles, and sustained seated or standing postures. It also calls for integral and more effective procedures for the assessment of ergonomic risks, which all companies should adopt, in order to identify and address the most relevant risk factors.

However, the assessment of ergonomic risks is often neglected or limited to the evaluation of critical safety risks (e.g. accidents and injuries), since the disorders caused by other ergonomic issues, like those related to postures, movements and manual handling, are normally developed only after a prolonged exposure to them, and often the relationship between daily tasks and those risks is not obvious. Another reason for such neglection, specially in small companies with limited resources, is the great time investment that is required to conduct an adequate, integral assessment of ergonomic risks [2].

There are methodologies like OWAS [3], RULA [4], REBA [5], etc., specially aimed at that type of ergonomic assessment, which provide ergonomic engineers and technicians with systematic criteria to assess a workplace, based on the analysis of observed postures. But comprehensive evaluations, specially for complex or combined tasks, require the observation, selection and analysis of postures from long working periods, which is time consuming and subject to bias.

Multiple tools have been developed to address such problems in the application of those methods, mostly based on automatic measurement of postures with wearable sensors [6, 7], depth cameras [8–10] or virtual reality scenarios [11, 12]. But although those solutions reduce the subjectivity of the assessments, and the time required by the evaluators to conduct the analysis, they increase the burden of taking the measurements, because workers have to be instrumented with devices that usually require a previous calibration, and the instrumentation or the setting of the scenario may also interfere with the task that is being measured.

In this work we present a new approach for the automatic assessment of ergonomic risks based on Artificial Vision and Neural Network methodologies, which addresses the two main problems that have been detected before: the subjectivity of the evaluator, and the time needed to perform those measurements. The efficacy of that tool has been assessed in terms of time needed for the assessment, and performance in the detection of postures for a specific use case based on the OWAS methodology.

2 Methods

A web application has been developed in PHP and Javascript to assist the assessment of ergonomic risks, with semi-automated image analysis to detect postures of different parts of the body. The application can be used in any device with a web browser and Internet connection, and includes functionalities to upload and pre-process videos in different standard formats (AVI, MPEG, MP4, etc.), and visualize and export the results of the evaluations.

The application allows to define tasks composed by an arbitrary set of subtasks to be analyzed, associating each subtask with a different fragment of one or several videos, and choosing a specific number of frames to be extracted from those fragments for each subtask. Selected video fragments are sent to an AWS Elastic Computing service, and automatically processed by a convolutional neural network (CNN) based on the Simple Pose deep learning network, a robust, fast and accurate CNN that has achieved very good results in recent benchmarks (AP of 73.7 on COCO dataset) [13], with custom end layers to calculate the parameters that are used in ergonomic workplace assessments.

This paper presents a use case for the evaluation of tasks based on the OWAS method [3]. The CNN, originally designed to return the coordinates of characteristic points of the face, trunk, arms and legs from people detected in each frame of the video, has been complemented with additional calculations that allow to classify the postures of trunk (straight, bent, and/or twisted), arms (below or above elbows) and legs (standing, sitting, kneeling, walking, and different combinations of knee flexion and foot support).

Those parameters are automatically calculated and stored in a database. A visual tool can be later used to review the detected postures, modify them, and add the load level that is needed to calculate the ergonomic risk of each frame (see Fig. 1).

To test the performance of the tool, it has been used with a sample of videos taken from the Carnegie Mellon University Motion Capture Database [14], particularly the samples #02.06 (bend over, coop up, rise, lift arm), #13.23 (sweep floor), and #15.06 (lean forward, reach for). The web application was used in a laptop with an Intel i5 2.3 GHz processor, 8 GB RAM and LAN Internet connection. The time for the analysis and the number of successful posture detections, considering the confidence level of joint coordinates computed by the Simple Pose CNN [13].

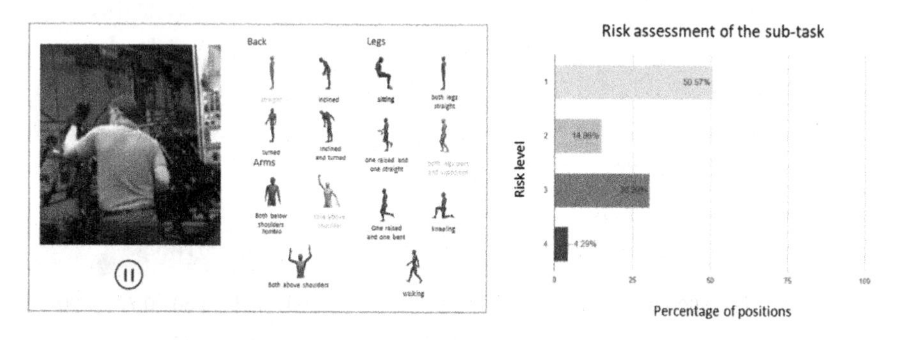

Fig. 1 Visualization and plot of results for OWAS

3 Results

The processing speed was between 15 and 20 frames per second, both in the pre-processing phase (trim and subsample) and in the analysis.

The CNN provided data of the postures of trunk, arms, legs in each frame. The confidence level for body part detections was over 0.5 (50%) in the majority of cases; and setting a minimum threshold of 25% of confidence, it was possible to assess the ergonomic risk based on the postures in between 73 and 86% of the images. The best results were obtained for legs, and the worst for arms, with a substantial decrease of performance in the sample video where the subject was kneeling down (#02.06, see Fig. 2).

Visual inspection of the videos with superimposed wireframes of joint coordinates (Fig. 3) showed that the poorer results in sample #02.06 were due to the

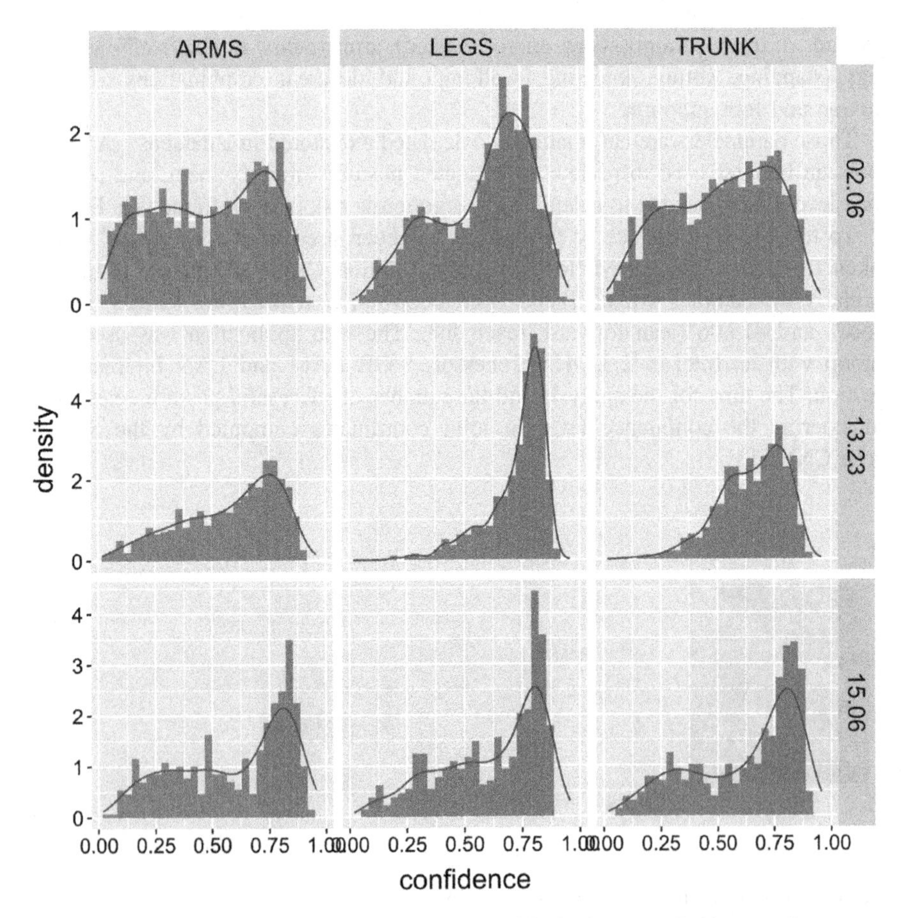

Fig. 2 Distributions of the confidence levels, separated by body part and video sample

Fig. 3 Example of image with superimposed wireframe (sample video #13.23), from [14]

subject's extreme crouched posture in approximately 20% of the frames. The following reasons for potential failure of posture detection were detected:

- Confusions of laterality (flipped left and right sides of the body). This happened more frequently in the sample #13.23, presumably because the homogeneous dark clothing of the subject and his partial face covering made the body recognition more difficult. However, this did not affect the results of the OWAS assessment, which is insensitive to laterality.
- Failure in body shape recognition. More frequent in unusual body postures (e.g. the crouched posture in the sample #02.06).
- Occlusion of body parts. This happened also in #02.06, which was close to a side view, and resulted in the hidden arm being assimilated to the other one.

4 Discussion

With the processing speed obtained in the tests, a one-minute video recorded at a standard frame rate of 30 frames per second can be fully processed in two less than two-minutes; or a 30 min video, subsampled at 100 frames per minute (with a total amount of 3,000 images), could be processed in 6 min.

The occasional occurrence of failures in posture recognition still enforces a revision of the results from the evaluator's side. Given the underlying reasons of the observed failures, some of them (e.g. flipped left/hand sides, unusual postures) are expected to be improved with further training of the CNN, using more images from workplaces, and specializing them further to give the results needed for the ergonomic assessment methods.

The approach presented for the use case of OWAS can be extended to other methods like RULA or REBA, by just modifying the final calculations obtained by Artificial Intelligence, to provide the postural parameters used by those methods. Thus, this tool is expected to facilitate more exhaustive and objective evaluations of ergonomic risks using postural assessment methods, and help to reduce the incidence of MSKD in work places, with faster times and smaller costs.

5 Conclusion

This application saves time compared with traditional procedures based on visual inspection, specially for great amounts of images, which nowadays is only matched by instrumented methods. The advantage of our approach is that it does not require the worker to wear any sensors, special cameras, or calibration of the work space, further reducing time and material investments.

But even considering its current performance, the time spent in the selection of images and labelling postures is smaller than with the traditional, manual approach; and this method also reduces the mental workload of evaluators and the dependency on their expertise. Thus, the same technician can evaluate a greater number of work places or analyze the same place for different workers, as recommended by the EU-OSHA [1].

Acknowledgements The data used in this project was obtained from the CMU Motion Capture Database (http://mocap.cs.cmu.edu), created with funding from NSF EIA-0196217.

References

1. Work-related MSDs: prevalence, costs and demographics in the EU—Summary 2019, https://osha.europa.eu/es/publications/msds-facts-and-figures-overview-prevalence-costs-and-demographics-msds-europe/view. Accessed 16 Oct 2020
2. Westgaard RH, Winkel J (1997) Ergonomic intervention research for improved musculoskeletal health: a critical review. Int J Ind Ergon 20(6):463–500
3. Karhu O, Kansi P, Kuorinka I (1977) Correcting working postures in industry: a practical method for analysis. Appl Ergon 8(4):199–201
4. McAtamney L, Corlett EN (1993) RULA: a survey method for the investigation of work-related upper limb disorders. Appl Ergon 24(2):91–99
5. Hignett S, McAtamney L (2000) Rapid entire body assessment (REBA). Appl Ergon 31(2):201–205

6. Lu ML, Barim MS, Feng S, Hughes G, Hayden M, Werren D (2020) Development of a wearable IMU system for automatically assessing lifting risk factors. In: International conference on human-computer interaction. Springer, Cham, pp 194–213
7. Nath ND, Chaspari T, Behzadan AH (2018) Automated ergonomic risk monitoring using body-mounted sensors and machine learning. Adv Eng Inform 38:514–526
8. Manghisi VM, Uva AE, Fiorentino M, Bevilacqua V, Trotta GF, Monno G (2017) Real time RULA assessment using Kinect v2 sensor. Appl Ergon 65:481–491
9. Otto M, Lampen E, Auris F, Gaisbauer F, Rukzio E (2019) Applicability evaluation of Kinect for EAWS ergonomic assessments. Procedia CIRP 81:781–784
10. Plantard P, Shum HP, Le Pierres AS, Multon F (2017) Validation of an ergonomic assessment method using Kinect data in real workplace conditions. Appl Ergon 65:562–569
11. Li X, Han S, Gül M, Al-Hussein M, El-Rich M (2018) 3D visualization-based ergonomic risk assessment and work modification framework and its validation for a lifting task. J Constr Engg Manage 144(1):04017093
12. Rizzuto MA, Sonne MW, Vignais N, Keir PJ (2019) Evaluation of a virtual reality head mounted display as a tool for posture assessment in digital human modelling software. Appl Ergon 79:1–8
13. Li J, Su W, Wang Z (2020) Simple pose: rethinking and improving a bottom-up approach for multi-person pose estimation. In: Proceedings of the AAAI conference on artificial intelligence, vol 34(07)
14. CMU Graphics Lab Motion Capture Database, http://mocap.cs.cmu.edu. Accessed 19 Oct 2020

Space, Time and Ergonomic Assessment of Order Picking Using Horizontal Carousel

Goran Đukić, Tihomir Opetuk, and Brigita Gajšek

Abstract The efficiency of order picking operations can be enhanced by different automated storage/retrieval systems (AS/RS) based on the "part-to-picker" principle. One such system is a horizontal carousel, a rotating rack that delivers parts to the operator. Apart from space and time reduction related benefits, horizontal carousels are often marketed as well as better ergonomic systems compared to traditional shelving rack systems. However, not many papers deal with a quantitative comparison of horizontal carousels with shelving systems that are much more used for small parts order picking. Especially there is a lack of papers analyzing ergonomics of horizontal carousels. In this paper, horizontal carousels are briefly presented, followed by methods to compare horizontal carousels with shelving systems in terms of space, time and ergonomics. For ergonomic assessment, a method of energy expenditure calculation is used. Presented methods can serve as tools for managers for additional justification of horizontal carousel's implementation.

Keywords Order-picking · Vertical lift module (VLM) · Ergonomic assessment · Energy expenditure

1 Introduction

Automated storage and retrieval systems (AS/RS) now have a long history of application in distribution and manufacturing environments. Dating back to the 1960 and the appearance of the first unit-load AS/RS, various types of AS/RS developed since then, capable of storing and retrieving materials (loads) of different

G. Đukić (✉) · T. Opetuk
Faculty of Mechanical Engineering and Naval Architecture, University of Zagreb, Zagreb, Croatia
e-mail: goran.dukic@fsb.hr

B. Gajšek
Faculty for Logistics, University of Maribor, Celje, Slovenia

D. Sumpor et al. (eds.), *Proceedings of the 8th International Ergonomics Conference*, Advances in Intelligent Systems and Computing 1313,
https://doi.org/10.1007/978-3-030-66937-9_9

sizes and weights. In many cases, the advantages of AS/RSs over non-automated systems are the reason why such systems are used for order-picking. As one of the four main warehousing processes (receiving, storage, order-picking and shipping), the order-picking process is the most labor-intensive and the costliest activity in a typical warehouse. According to the [1], up to 55% of warehouse total operating costs are attributed to order-picking. Additionally, most incorrect shipments due to the wrong amount and/or type of ordered items could be addressed to the picking process. It is obvious why many companies are improving their order-picking operations by using more efficient systems. In classical order-picking systems (usually from the shelf and standard pallet racking systems), pickers' traveling amounts to around 50% of the total picking time [1], so the logical way of improving this is to reduce or eliminate unproductive walking time. The efficiency of order-picking operations can be enhanced using AS/RSs.

As said, there are various types of AS/RSs. Although some consider AS/RS usually systems with automated cranes or S/R machines (and are called crane-based AS/RSs), like in [2], nowadays various other systems are also automated storage/ retrieval systems. With a little bit longer history, there are horizontal and vertical carousels and vertical lift modules (VLM), while the newest types are systems with racking vehicles (shuttles), called Shuttle-based S/RS (SB-S/RS) or Autonomous Vehicles S/RS (AVS/RS). Although all types of those systems have applications in order-picking processes, rarely all benefits are quantified and evaluated.

The order-picking process is, in general, very labor-intensive in which workers perform ergonomically risky postures of bending, reaching and lifting items manually. However, research papers about ergonomics (human factors) in order-picking processes are rare and have appeared only recently. Moreover, they focus mainly on manual order-picking systems, i.e. "picker-to-part" systems where picking is done from static systems like shelving or pallet racks, with walking or traveling between picking locations. Except in [3], no papers present ergonomic quantitative comparison of AS/RS system with shelving systems as their much more frequently used counterpart.

This paper focuses on a horizontal carousel, a rotating rack that delivers parts to the operator (picker). Considering this, potential users might be initially attracted to it because of increased productivity as a result of their main advantage, i.e. the operator (picker) stays in one particular location and items are brought to the operator by the system. Hence, these systems are termed "part-to-picker", „stock-to-operator" or „end-of-aisle" systems. As pickers pick from one position, there is no need for aisles along this rack, so another expected benefit is also better space utilization. There are also other less visible benefits, like improved ergonomics, improved picking accuracy (fewer picking errors), greater security (less theft and damage of stored items) and better inventory control (improved inventory accuracy). In this paper, the ergonomics assessment of such systems for order-picking, along with space and time, is the primary goal.

The structure of the paper is as follows. In the second chapter, horizontal carousels are explained, followed by presented methods used to compare horizontal carousel with the shelving system based on one selected example, to provide quantification of possible differences (improvements achieved when using horizontal carousels). The presented methods can serve managers as tools to justify the implementation of horizontal carousels additionally.

2 Horizontal Carousels

There is a vast number of papers dealing with AS/RSs. Most of them are related to various crane-based systems, mostly unit-load AS/RSs and mini-load AS/RSs. The most comprehensive overview of the design and control issues of such AS/RSs is given in [2]. Lately, a significant number of papers appeared regarding the design and control of SBS/RSs, of which we point as well to a review paper [4]. Horizontal and vertical carousels received much less attention in the literature. Papers mostly regard operating policies and design, of which we point on [5]. In that paper, the authors referenced other older papers about carousels, dealing mostly with retrieval sequencing, item assignment policies and carousel/s throughput with robotic interfaces. VLMs received the least attention, just lately several papers regarding throughput models and operating strategies were published, for instance [6, 7].

According to the definition given by MHI, a horizontal carousel is a storage device that consists of a fixed number of adjacent storage columns or bays that are mechanically linked to either an overhead or floor mounted drive mechanism to form a complete loop. Each column is divided into a fixed number of storage locations or bins, which in most applications are constructed of a welded wireframe, as illustrated in Fig. 1.

Loads consisting of containers or totes may be inserted and retrieved manually or by an automatic inserter/extractor mechanism. However, the carousel rotation, whereby a specific storage location is brought to the picking location, is almost always controlled automatically.

Horizontal carousels have been in use since the early 1960s, only later became widely used for small parts storage integrating PCs and basic hardware devices. They can be used for primary storage, buffer storage, order-picking, staging, consolidation and sortation [9]. Therefore, basic applications tended to be primarily in storage applications in manufacturing environments rather than order-picking applications, utilizing a horizontal carousel as a storage machine. Other applications of horizontal carousels nowadays are as components of distribution systems for order-picking operations. With the primary purpose of achieving high picking throughputs, today's horizontal carousels are found usually in pods, integrated work centers configurations of 2, 3 or even 4 horizontal carousels (units) per operator (human picker). Figure 2 illustrates one such example.

The benefits of using horizontal carousels are numerous. When the first horizontal carousels were used as stand-alone machines, the fact that parts are brought to the operator instead of need to walk to the parts enabled operators to spend more time on other work (paperwork, counting, etc.). In pods, where operators handle more carousels at once, waiting time (wasted time) is further reduced or eliminated,

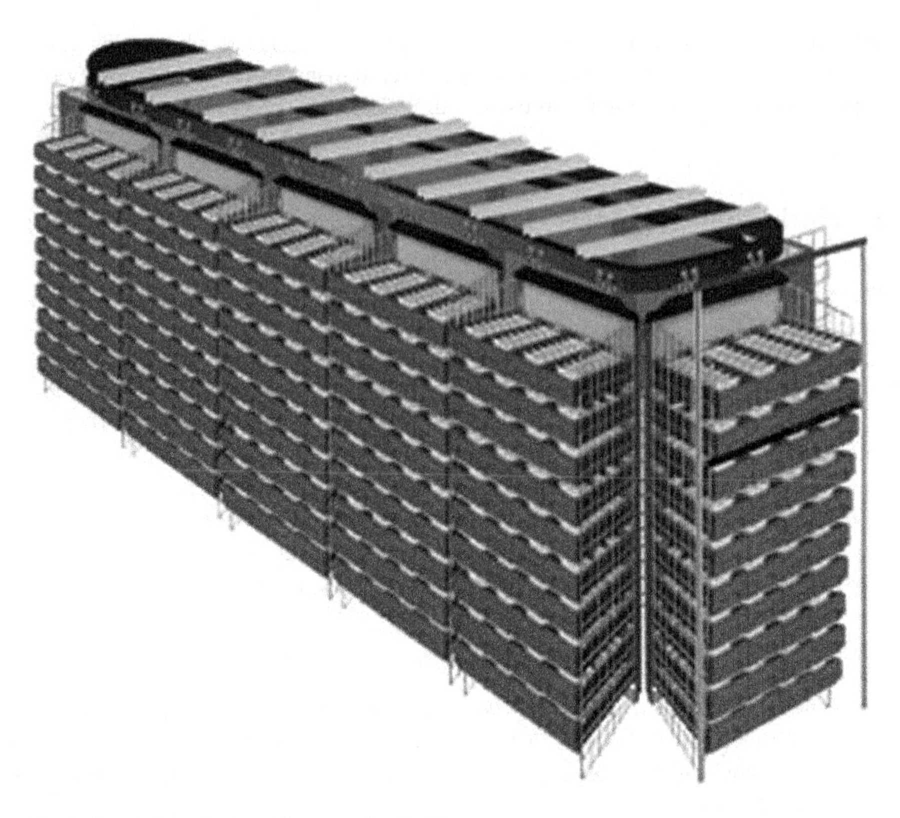

Fig. 1 Stand-alone horizontal carousel unit [8]

enabling high picking productivity. The fact that picking from the carousel is done from one position (one end of the unit), there is no need to access from the side (as in shelving systems). Eliminating the need for aisles between units saves floor space. Space benefits could be further achieved with stacked carousels (higher units), with a mezzanine's need only to support the operator (not on the entire storage area). Better storage space usage is also achieved with better storage density, resulting from the minimal spacing needed between storage bins. This will be illustrated in the third chapter. Horizontal carousels equipped with computers and advanced software are distinguished by inventory management (location, inventory control). The fact that they are accessible only from one side is reflected in better security of stored items on the one hand and increased control of workers by supervisors. And finally, the elimination of walking reduces the operator's fatigue and boredom. Apart from being more ergonomic, it will also reduce inaccuracies caused by fatigue. Altogether, it leads to better productivity and better accuracy.

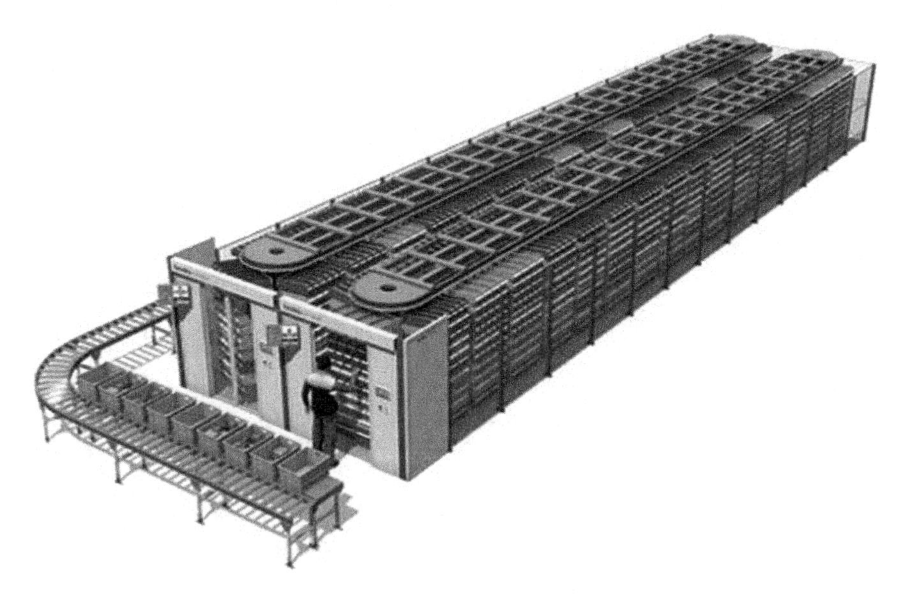

Fig. 2 Two horizontal carousel units in pod [10]

3 Comparison of Horizontal Carousel and Shelving System

To justify horizontal carousels in comparison with shelving systems, one should try to quantify their essential benefits (decreased storage space requirement, decreased picking time) along with other benefits (in this paper, the focus is on ergonomics). In this paper, preliminary research results are presented, with analysis limited to only one stand-alone horizontal carousel assumed for comparison. In future research, analysis will focus on 2-unit and 3-unit carousel pod configurations as well.

For the purpose of quantitative assessment, one example was selected, shown in Fig. 3 on the left side. The shelving area with 8 shelves 1.74 m high (5 levels with 0.348 m height) has a total of 74.12 m^3 of storage space.

Given the usually relatively low space utilization of shelves (according to the [11] utilization of shelving was selected as 30%), the total required storage space for stored items is 22.185 m^3. Again, given the usually better utilization of storage space in horizontal carousels (according to the [11] utilization of horizontal carousel was selected as 62%), the total required storage space in horizontal carousel amounts to 35.78 m^3. For easier and more direct comparison, a horizontal carousel is assumed with the same number of levels, height of bins and depth of bins as in shelves (therefore 5 levels with 0.348 m height, 0.5 m depth). The width of the bins (therefore columns or carriers) was selected as 0.622 m (as the actual dimension of one particular model of well-known producer). Dimensions of the carriers (bins)

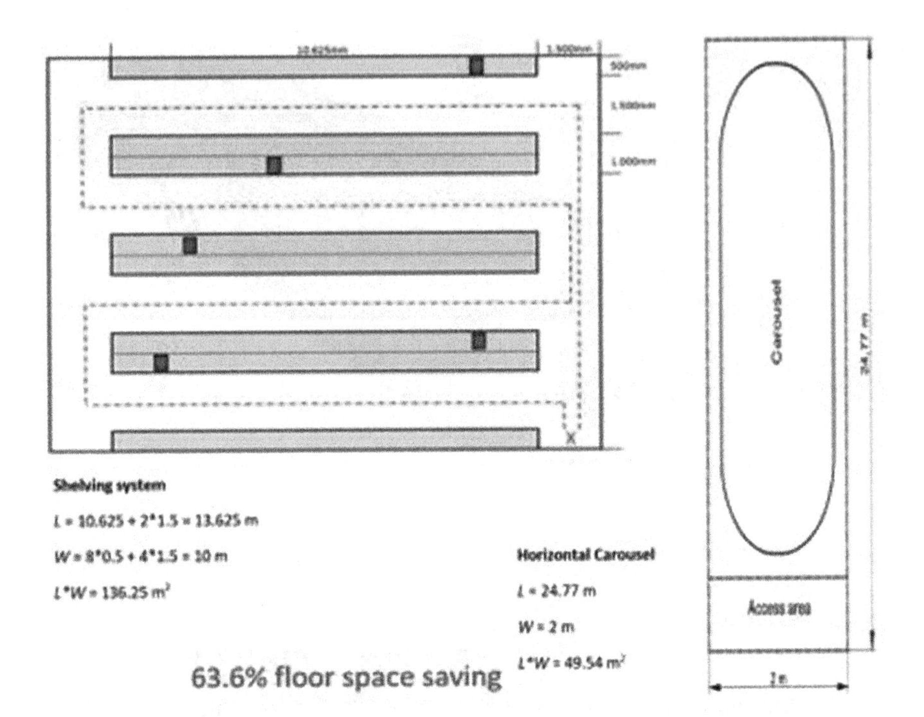

Fig. 3 Space assessment comparison and picking route illustration

together with the required space led to the required number of carriers $n_c = 66$. For such configuration and dimensions, the required floor space for one horizontal carousel is shown in Fig. 3 on the right side. Comparing the required total spaces, the floor space-saving is more than 60%.

Assessment of time for picking was done, assuming picking orders with several different items (stock keeping units, SKUs) per order. Two different order sizes were analyzed—one assuming small orders with 5 picks per order and another assuming larger orders with 30 picks per order.

As already said, picking from the manual "picker-to-part" system requires walking from pick location to pick location, while picking from horizontal carousel is "part-to-picker" system and bins (with items) are mechanically delivered to the picker. However, while picking from shelves requires walking, picking from a horizontal carousel requires waiting for delivery of the next bin (rotated to the picking place). Assuming that random storage policy is used, there are the same probabilities for picking from levels (therefore for stretching, bending or picking from the workers' waist level) of the carousel as it is from shelves. Although in shelving systems might be unneglectable search time (time to find shelf location, check location and item) while carousel system will bring proper location, we will assume exactly the same time needed for picking per item (for shelving system equals search time plus retrieval time, in horizontal carousel system equals stop/start time plus search time plus retrieval time). For this example, picking time t_p was selected as 10 s.

The picking time of one order in manual "picker-to-part" systems greatly depends on the routing policy used. Routing policies define the sequence of the picking and actual picker's walking route. There are plenty heuristic policies developed and used, as well as an optimal algorithm [12]. In this research, we assumed a simple yet most popular policy, called S-shape (traversing all aisles in a serpentine way), as illustrated with the route in Fig. 3 on the left side. Travel time was calculated using the model presented in [13], assuming acceleration 1 m/s^2.

Retrieval sequence greatly influences the total traveling time of the carousel as well. The simplest model assumes retrieval of each item as a single order, defining retrieval time as

$$T_r = C/(0.25 * v_c) + t_p \quad (1)$$

where C is the circumference of the carousel track (in this example calculated as 45.5 m), v_c is carousel velocity (in this example taken 24 m/min, as actual speed of one particular model of well-known producer), while t_p is already defined picking time per item. Simple calculations show that picking times for an entire order of 5 and 30 items ($5 * T_r$ and $30 * T_r$, respectively) are significantly larger than the picking from the shelves (see Table 1). Waiting times with one carousel under such operational policy is too high, making use of one carousel not time beneficial.

However, rotation time is greatly reduced with sequencing. Assuming it, the authors in [5] developed the model that calculates expected total retrieval and picking time $E^C(n)$ of an order as a sum of expected rotation time $E^R(n)$, expected start/stop time $E^S(n)$ and expected picking time $E^P(n)$. In our research, we slightly simplified this model by assuming that carriers' start/stop time is included in picking time per item t_p. Therefore, our total time to pick an order is calculated as

$$E^C(n) = E^R(n) + n * t_p \quad (2)$$

Table 1 Time assessment comparison

Parameters and results/system and model	Shelving system, S-shape strategy		Horizontal carousel, w/out sequencing		Horizontal carousel, with sequencing	
n (number of items to pick per order/batch)	5	30	5	30	5	30
t_p (time to pick one item), s	10	10	10	10	10	10
Expected time for travel/rotation, s	84.5	109.5	142.2	853.2	86.5	108.3
Expected time for picking, s	50	300	50	300	50	300
Expected total time for picking an order, s	**134.5**	**409.5**	**192.2**	**1153.2**	**136.5**	**408.3**

where $E^R(n)$ is calculated using the proposed model in [5], for n equals 5 and 30, respectively.

Results in Table 1 show significantly shorter total retrieval times with sequencing compared to the model without sequencing. Additionally, the total time to pick an order from the carousel with sequencing is practically the same as the total time to pick the same order from the shelves. This finding put time (productivity) benefit in question. Waiting time for the next bin is similar to travel time between locations in shelving system (while pick time per item was assumed to be equal), so no time reduction is achieved. One should be careful about this because usually marketed time benefits of carousels consider systems with units in a pod.

The human plays a key role in order-picking systems. Integrating the ergonomics aspect into the workplace design is not a new topic (many papers are dealing with ergonomics in manufacturing and assembly environments), also recognized in practice, in internal logistics and manual material handling. In manual "picker-to-part" systems, activities require manual material handling tasks, such as lifting, lowering and carrying loads, while usually associated with awkward postures, such as reaching and bending. Order-picking activities involve repetitive tasks that may result in musculoskeletal disorders (MSDs) for workers. In "part-to-picker" systems, the use of AS/RSs eliminates walking, and some systems eliminate also reaching and bending (enabling picking from the most ergonomic position, so-called "golden zone"). However, scientific literature on ergonomic aspects in order picking is scarce and appears only in the last few years. This was recognized in [14], so the authors developed a conceptual framework that identifies steps in an order-picking involving humans. This research was followed and supported by [15]. The authors performed the content analysis of the literature about the human factor in order picking, concluding that the focus of the literature on manual order picking has mainly been on economic aspects and that human factor issues have not been mentioned very often. Although the applications of AS/RSs are aimed to reduce walking distance and improve ergonomics due to reduced walking and carrying items and picking from the most ergonomic position, no papers were particularly devoted to this issue except the paper for ergonomic assessment of VLM in [3].

For the ergonomic assessment of the horizontal carousel, the calculation of energy expenditure was used based on the equations and values given in [16], influenced by the manual order-picking system analysis in [17]. Items to be picked were of different mass and volume, randomly stored in shelf rack (this situation was analyzed in one laboratory experiment for analysis of picking time, results were presented at a conference, and published in a journal). For those 9 different combinations, average energy expenditure per pick was calculated using equations in [16, 17] as $E_p = 0.795$ kcal/pick, energy expenditure for walking as $E_w = 0.0806$ kcal/s and energy expenditure per standing as $E_s = 0.264$ kcal/s.

The energy expenditure for picking an order from shelves depends on the energy used for walking and energy used for picking from different levels (bending, reaching) while standing. In the case of picking from a horizontal carousel, energy expenditure consists of energy for standing during waiting and energy used for

Table 2 Ergonomic assessment comparison

Parameters and results/system and model	Shelving system, S-shape strategy		Horizontal carousel, with sequencing	
Energy for picking, kcal	5 * 0.795 = 3.975	30 * 0.795 = 23.85	5 * 0.795 = 3.975	30 * 0.795 = 23.85
Energy for walking, kcal	84.5 * 0.0806 = 6.8107	109.5 * 0.0806 = 8.8257	–	–
Energy for standing during picking, kcal	5 * 10 * 0.0264 = 1.32	30 * 10 * 0.0264 = 7.92	5 * 10 * 0.0264 = 1.32	30 * 10 * 0.0264 = 7.92
Energy standing during waiting, kcal	–	–	86.5 * 0.0264 = 2.2836	108.3 * 0.0264 = 2.8591
Total energy per order, kcal/order	**12.1057**	**40.5957**	**7.5606**	**34.6291**
Energy per picked item, kcal/picked item	**2.421**	**1.3532**	**1.5121**	**1.154**

picking from different levels (again bending, reaching) while standing. The results (average energy used per order and per picked item) are presented in Table 2.

As could be seen from the results, the energy required for picking from the horizontal carousel is smaller than the energy required for picking from shelves. In this example, the total time to pick an order is similar in both cases (because waiting times with the carousel is similar to walking times in the shelving system), therefore influencing the result with the fact that energy used for standing is less than the energy used for walking.

4 Conclusion

In this paper, possible methods for assessing horizontal carousels are presented, focusing on space, time and ergonomic assessment.

Space assessment was done in a usual, logical way to compare the required space for the horizontal carousel(s) compared to the storage space needed for the shelving system to store the same inventory amount. The method is based on some empirical utilization data, provided in [11].

Time assessment of the carousel is more complex and depends on many more assumptions. The first one is that the horizontal carousel is used as a stand-alone machine with random storage policy and sequencing. In a shelving system, an S-shape routing policy and random storage are being performed. Another assumption was that pick times per item for both systems are identical. Stop/start time of carousel was simplified and combined with picking time (while assumed same time as search time in shelving system). Different assumptions, especially 2-unit or 3-unit carousels in pods, would significantly change resulting total picking times.

An ergonomic assessment is also complex. Order-picking items during one cycle usually involve different items (mass, volume) with different picking frequencies, picking from different levels, where the application of risk assessment methods for ergonomic analysis is not convenient. We followed the idea to assess this process using calculated energy. Still, different items (mass, volume) and different levels are also causing different energy expenditure values. In this example, we have chosen one average value (based on one other research) however, any other situation might result in different values. Therefore, in this case, we do not point on actual values, rather to the relative relationship—less energy for standing while waiting compared to the energy for walking. This approach also leads to the conclusion that in 2-unit or 3-unit carousels in pods, where waiting time could be eliminated, energy used per order will further decrease. However, this results in more orders picked per hour due to the increased productivity, leading to more energy for picking spend per hour. In future research, described challenges will be addressed.

References

1. Tompkins JA et al (1996) Facilities planning, 2nd edn. John Wiley & Sons Inc, Hoboken, NJ
2. Roodbergen KJ, Iris FAV (2009) A survey of literature on automated storage and retrieval systems. Eur J Oper Res 194:343–362
3. Đukić G, Rose L, Gajšek B, Opetuk T, Cajner H (2018) Space, time and ergonomic assessment of order-picking using vertical lift modules. In ICIL 2018 conference proceedings, Ben-Gurion University, Israel
4. Kosanic N, Milojevic G, Zrnic N (2018) A survey of literature on shuttle-based storage and retrieval systems. FME Trans 46(3):400–409
5. Meller RD, Klote JF (2004) A throughput model for carousel/VLM pods. IIE Trans 36:725–741
6. Dukic G, Opetuk T, Lerher T (2015) A throughput model for a dual-tray Vertical Lift Module with a human order-picker. Int J Prod Econ 170:874–881
7. Calzavara M, Persona A, Sgarbossa F (2018) Economic and performance analysis of dual-bay Vertical Lift Modules. In 15th international material handling research colloquium (IMHRC 2018), Savannah, USA
8. https://www.mhi.org/solutions-community/solutions-guide/horizontal-carousels
9. Frye R, Roche C (2003) Dynamic storage systems: horizontal carousels. In The essentials of material handling: part 2—the basic product knowledge program. Material Handling Industry of America, Charlotte, North Carolina
10. https://www.kardex-remstar.co.uk/uk/storage-retrieval-systems-uk/horizontal-carousels/stations.html
11. ASRS capacity & floor space savings. Kardex Remstar, white paper
12. De Koster R, Le-Duc T, Roodbergen KJ (2007) Design and control of warehouse order picking: a literature review. Eur J Oper Res 182(2), 481–501
13. Ten Hompel M et al (2007) Materialflusssysteme. Springer-Verlag, Berlin Heidelberg
14. Grosse EH, Glock CH, Yaber MY, Neumann P (2015) Incorporating human factors in order picking planning models: framework and research opportunities. Int J Prod Res 53(3):695–717
15. Grosse EH et al (2017) Human factors in order picking: a content analysis of the literature. Int J Prod Res 55(5):1260–1276
16. Garg A et al (1978) Prediction of metabolic rates for manual materials handling jobs. Am Ind Hyg Assoc J 39(8):661–674
17. Battini D, Glock CH, Grosse EH, Persona A, Scarbossa F (2016) Human energy expenditure in order picking storage assignment: a bi-objective method. Comput Ind Eng 94:147–157

References

1. [illegible] 2nd edn. John Wiley & Sons Inc, Hoboken, NJ
2. [illegible]
3. [illegible]
4. [illegible]
5. [illegible]
6. [illegible]
7. [illegible]
8. [illegible]
9. [illegible]
10. [illegible]
11. [illegible]
12. [illegible]
13. [illegible]
14. [illegible]

Modified NIOSH Lifting Equation with MATLAB

Gyula Szabó

Abstract Work-related musculoskeletal diseases despite decades of efforts, represent a leading workplace risk. The NIOSH modified lifting equation, available in various forms and platform, can be used to assess the risk from manual material handling. While assessing just a few lifts per shift or homogeneous lifting situations is relatively simple; however, in a mixed, high-frequency lifting situation, the risk assessment involves significant data collection and requires complex data processing. In this paper, we present the MATLAB processing of manual handling data retrieved from a production program through an assembly line job as an example. We show the transfer functions for each risk factor and the importance of accurate input data and evaluate their impact on the ultimate risk.

Keywords Ergonomics · Manual lifting · Workplace health and safety · Modified lifting equation

1 Introduction

According to the European Working Conditions Survey, manual handling is a leading occupational hazard, and more and more workers are affected [1]. Musculoskeletal disorders are also a significant hazard in the workplace, reported the latest EU-OSHA survey [2].

Reducing the health effects of manual handling in different settings can be achieved according to the principles of risk prevention. Automation, machine material handling can mean the elimination of hazards, up to individual-dependent solutions such as training workers to perform manual material handling operations correctly or using exoskeletons.

The European Agency for Safety and Health at Work's current Lighten the load campaign aims to reduce the risk of work-related musculoskeletal disorders. The

G. Szabó (✉)
Óbuda University, Budapest, Hungary
e-mail: szabo.gyula@uni-obuda.hu

D. Sumpor et al. (eds.), *Proceedings of the 8th International Ergonomics Conference*, Advances in Intelligent Systems and Computing 1313,
https://doi.org/10.1007/978-3-030-66937-9_10

campaign also proposes innovative solutions and previously neglected solutions, such as occupational safety for those returning to work after illness, or the role of educators in preparing children to withstand physical exertion, and implementing various health-improving measures, such as workplace physiotherapy solutions.

2 Risk Assessment of Manual Handling of Loads

Factors influencing the risk of manual handling have long been understood, and accepted assessment methods are available.

The numerical incorporation of interactions between psychosocial risks and the risk of movements into risk assessment awaits itself. Several of the theoretical contexts are already known, for example, in the case of welders, quality welding under time constraints is difficult to harmonise with the correct posture, i.e. it is almost impossible to weld comfortably, much in good quality.

The generally accepted method for risk assessment of manual lifting is the NIOSH Modified Lifting Equation [3]. The risk factors included in the assessment are also reflected in European legislation [4], and this method is included in EN 1005-2 [5].

The modified lifting equation can be used with simplifications or in its original form. Some applications use a simplified form, e.g. Identification of critical cases described in EN 1005-2, CERA Manual Material Handling Sheet [6] or Manual Material Handling Assessment Sheet [7]. As paper-and-pencil methods, these allow the risk assessment of individual series of lifts consisting of roughly identical masses for a defined period, either with simple tables or checklists.

Although a modified lifting equation allows this in principle, the originally developed tabular estimation method is challenging to use for complex lifts. These make situations where the operator lifts varied weights during shorter or longer lifting periods separated by some rest time or a completely different activity.

In recent years, NIOSH's modified lifting equation has been incorporated into several computer-aided ergonomic evaluation systems or implemented in stand-alone applications, resulting in easy-to-use solutions such as the NIOSH application [8], ergo/IBV [9], or ErgoFellow [10].

These tools still leave it to the evaluator to identify a lift, define which period is considered a series of lifts, and how long the manual material handling takes. Similarly, input data is the weight of the load, the grip of the object, the marking of the team manual handling or environmental factors, e.g. floor surface, temperature.

3 Data from Logistics

In practice, a Logistics System contains information on the volume and nature of manual material handling, and this allows the evaluation of manual handling. In a survey, Vitos analysed the logistics process of a retail store. She identified four types of homogeneous lifting, i.e. bulky goods moved by machinery, heavy goods moved by hand, goods shelved by medium weight, and suspended goods in large quantities. Based on the stock movement, she was able to assess the workload of the workers and determine the risk of manual material handling. Consequently, among the measures developed following the principles of risk prevention, the direct delivery of the goods, from the distribution centre to the customer, without delivery to the store, was in the first place [11].

Analysing the material handling data of an automotive logistics service provider, Kovács determined the risk of manual material handling for a loading job. Since a computer system controlled the material movement, it registered all the lifts in the log. The computer system contained the IDs of the shelves and packages, dates and times of each movement, and the weights. It the field it was difficult to determine the location of the weight on the trolley or the original height of the incoming goods on the pallet. Kovács found the evaluation complication because the angle of asymmetry was challenging to determine [12].

4 NIOSH Transfer Functions

Using MATLAB, we constructed the transfer functions for each factor in the NIOSH modified lifting equation. By showing the quantifiable effect of risk factors, we can determine the accuracy of establishing individual data when designing machine data collection. At the same time, these transfer functions also shed light on which risk factors require the most attention when planning manual material handling and creating workplaces.

4.1 Expert Inputs

For applications that assess the risk of manual material handling based on automated data collection, certain factors also need to be determined by expert judgment, starting with the determination of the population-dependent reference mass. The reference mass varies over an extensive range, due to the variability of human force exertion, the acceptable 15 kg for the general working population can be up to twice as high for the unique working population.

The expert should evaluate several risk factors, e.g. floor quality, temperature. Similarly, the assessor should determine whether the operator lifts with one or two hands, the material handling situation is individual or group, whether there's an additional task between lifts or the quality of the grip.

According to NIOSH's modified lifting equation, these factors significantly reduce the recommended lifting weight limit. Depending on the expert assessment, the reduction is 40% for one-handed lift, 5–10% for poor grip, 20% due to the additional task and up to 15% for team lifting per person. Overall, based on expert judgments, the recommended weight limit can be reduced to 36%.

4.2 Asymmetry Component

The angle of asymmetry characterizes the posture correctness and measures the rotation of the body from the plane of symmetry. With the increase of the angle of asymmetry decreases proportionally the asymmetry component as well as the recommended weight limit. At the maximum allowable backstroke (asymmetry 235°), the recommended weight to be lifted is 43% lower than when lifted in front of the body.

4.3 Vertical Component

The vertical positioning factor in the NIOSH lifting assessment is ideal at the height of 75 cm from the ground level and decreases in proportion to the deviation measured from it. For example, the recommended lifting weight limit is 30% less than the ideal value at the highest acceptable lifting height (175 cm).

4.4 Distance Component

The vertical movement of the load is described by the Distance component, which is 100% up to 25 cm and then decreases by 15% at the maximum acceptable level difference (175 cm). Figure 1 shows the transfer function of the vertical load displacement, where right above 25 cm the factor decreases significantly, so a small deviation from the ideal cause a massive change in the factor; while among the already large load displacements, the Distance component no longer changes significantly. Consequently, in terms of computer processing, the data collection on load movement needs to be done more accurately for relatively small level differences. In contrast, for more extended level differences (100 cm) the risk of manual lifting is almost the same.

4.5 Horizontal Component

Figure 2 shows the transfer function of the Horizontal component, meaning the horizontal distance in front of the body. Lifting is risk-free when pressing the weight to the body (0–25 cm) and is unacceptable for long forward reaches that also require tilting forward. When lifting with the arm fully extended in extreme body position, only 40% of the recommended weight limit is acceptable.

Due to the course of the curve, data acquisition must be performed more accurately in the case of a smaller forward extension of the arm. In contrast, less

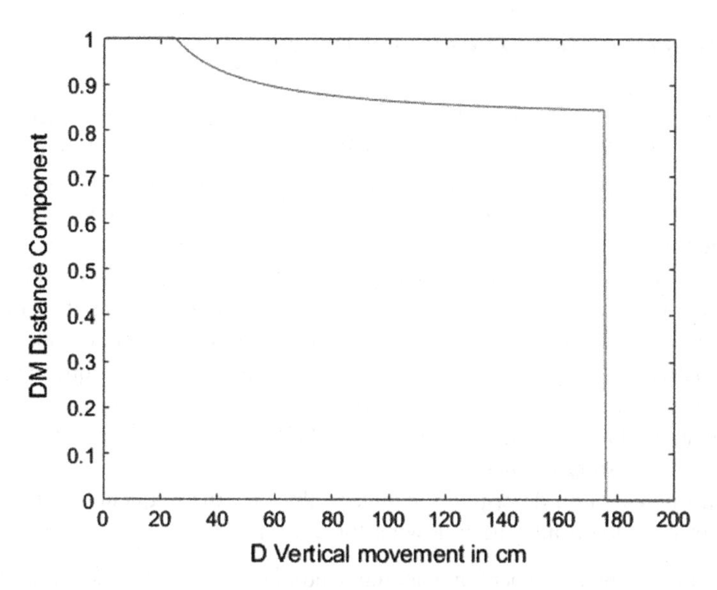

Fig. 1 The distance component transfer function

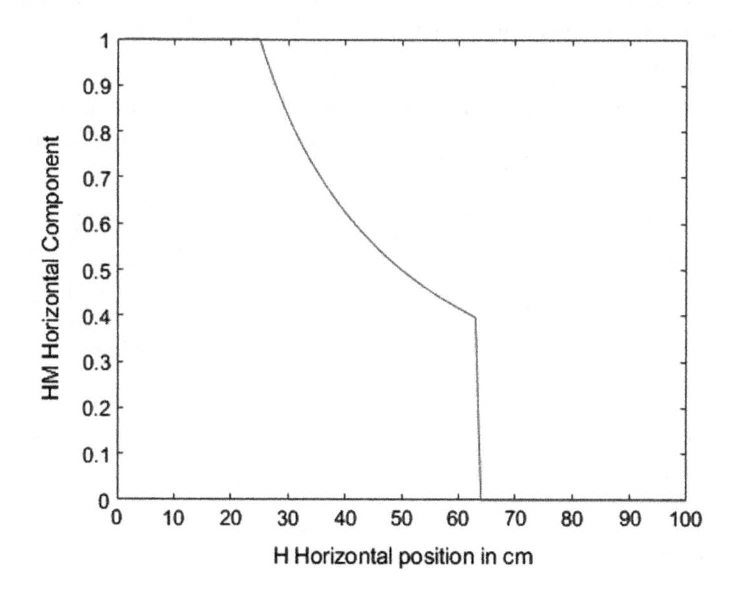

Fig. 2 The horizontal component transfer function

accurate data collection is sufficient in the case of far forward extensions. For example, with a forward reach 3 cm longer than the ideal, the reduction in lifting weight is already 10%, at 6 cm 20%, and at a double distance (50 cm) 50%.

4.6 Frequency (and Duration) Component

Not acceptable the manual lifting task if the weight exceeds the frequency-independent recommended lifting weight limit value. Experience has shown that the data obtained from the IT system provides enough information to determine the frequency-independent lifting index with the limitations and accuracy mentioned above.

The risk of manual lifting is also affected by the lifting frequency, and the time spent with manual handling, i.e. the total weight moved during the shift. The frequency (and duration) component represents the frequency and duration factors in the modified lifting equation allowing the assessment of lifting sequences, which are time-limited manual material movements after a significant break. Examples for such a material handling situation can be loading on the conveyor belt, or loading the goods at the store cashier.

The lifting equation description includes values for the frequency component in a table, broken down into the following three sections:

- according to the time spent on material handling from the shift: less than 1 h, for periods of 1–2 h, 2–8 h;
- frequency of lifts from one lift every two minutes, up to 15 lifts per minute; and
- for situations involving lifting heights under and over 75 cm.

Figure 3 shows the transfer function of the frequency component. The highest value is 1, which expresses that repetition for less than 1 h does not increase the risk of material movement if there are no more than two lifts per minute. This curve runs

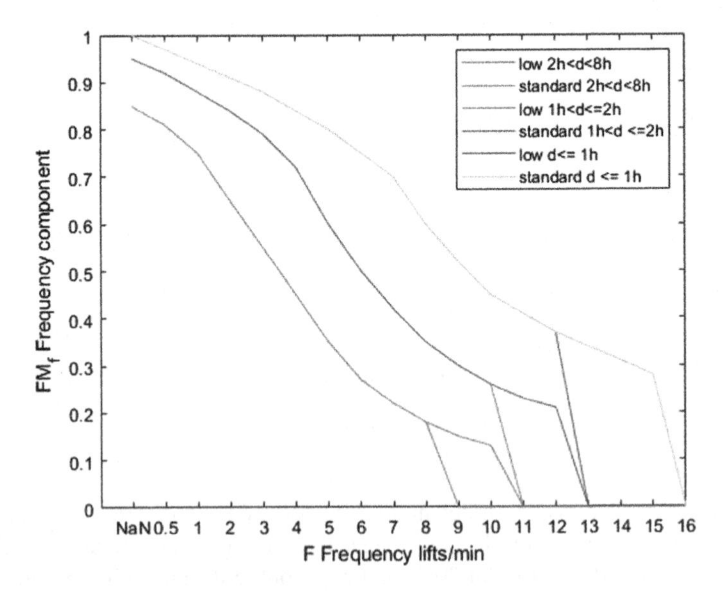

Fig. 3 The frequency (and duration) component transfer function

at the top and even for 15 lifts per minute includes that the recommended lift weight limit is only 28% of the initial repetition free frequency-independent value. The curve running below is the frequency component for material lifting for more than two hours, and even at a pace of one lift in two minutes, it is only 81%.

As the number of lifts per minute increases, the recommended weight value decreases significantly. On the right side of the curve, the curve branches off and decreases to zero for work under 75 cm, while it contains additional values for higher frequency lifts at standard height.

The frequency-dependent lifting index is the quotient of the average weight of the loads moved during the period assessed, and the frequency-dependent recommended lifting weight limit.

The lifting equation also allows determining the risk for the entire shift. Following the identification of lifting series in the shift, the lifting indices of the lifting series determined, and then the risk can be determined by weighting them together.

5 Results

The investigation revealed frequent torso bends and deep bends in all the area, with all employees assessed leading to increased risk of back disease and injury. Loading boxes full of parts over 10 kg onto a roller shelf significantly increased the risk due to the weight of the load and the frequency of lifting. Work with twisted wrist resulting from the gripping and rotation of the crates also contributed to increased risk of the musculoskeletal diseases on the upper limbs.

Excessive flexion of the upper arm and forearm occurred due to poor posture, primarily due to working over the shoulder and leaning deep. There was a significant risk of the forearm moving outside the 60°–110° angle range when moving KLT crates.

Recommendations for risk reduction based on the ergonomic risk assessment:

- Reducing the need for movement by optimizing the logistics system
- Transformation of lifting tasks, e.g. storing loads between 8–13 kg at the optimal height of 90–120 cm, the lighter ones can get higher or lower.
- Technical measures, e.g. lifting equipment, forklift truck system
- Organizational measures, e.g. regular breaks, rotation with assembly job
- Establishment of a selected physique working group for heavy lifting.
- Ergonomic development of the work area with the consultation of employees
- Training workers, mastering the correct lifting technique.

6 Summary

We dedicated this paper to the reduction of work-related musculoskeletal disorders resulted in manual handling of loads. The modified NIOSH lifting equation today is generally accepted and considered as a reliable method for determining the recommended lifting weight limit and evaluating the lifting index and the risk of manual lifting.

In this article, we demonstrated that computer-based logistics data collection could offer a way to assess the risk of manual material handling and to define preventive measures. The risk assessment requires, on the one hand, automated data acquisition and on the other hand, some expert judgments such as evaluation of the grip, two-handed or team lift.

It may be sufficient to consider only the initial or final position in the valuation if the lifting index does not eventually reach 59%, as this is the most substantial possible combined effect of the initial and final evaluations.

In the case of vertical distance and horizontal load placement, the lifting equation is the most sensitive to a slight deviation from the ideal. In the case of significant level differences or an even worse position of the already extended arm, the change is less reflected in the change in the recommended lifting weight limit.

Further research will determine how the breaking down of shifts into lifting series affects the complex lifting index. Hopefully, more research will reveal ways to calculate the complex lifting index without the assessment of lifting series only using frequency-independent lifting indices and the exact timing of each lift.

References

1. EUROFUND: European working conditions survey, https://www.eurofound.europa.eu/data/european-working-conditions-survey. Last accessed 3 Nov 2020
2. Crawford JO, Graveling R, Davis A, Giagloglou E, Fernandes M, Markowska A, Jones M, Fries-Tersch E (2020) Work-related musculoskeletal disorders: from research to practice. In What can be learnt? European Agency for Safety and Health at Work. ISBN: 978-92-9479-357-7 https://doi.org/10.2802/118327
3. Applications manual for the revised NIOSH lifting equation, January 1994, DHHS (NIOSH) Publication Number 94-110
4. Council Directive 90/269/EEC of 29 May 1990 on the minimum health and safety requirements for the manual handling of loads where there is a risk particularly of back injury to workers (fourth individual Directive within the meaning of Article 16(1) of Directive 89/391/EEC)
5. EN 1005-2:2003+A1:2008 safety of machinery—human physical performance—part 2: manual handling of machinery and component parts of machinery
6. Szabó G, Németh E (2019) Development an office ergonomic risk checklist: composite office ergonomic risk assessment (CERA Office) In Fujita Y, Alexander T, Albolino S, Tartaglia R, Bagnara S (eds) Proceedings of the 20th congress of the International Ergonomics Association (IEA 2018). Springer International Publishing, pp. 590–597, paper: Chap 64, 8 p

7. Manual handling assessment charts (the MAC tool) https://www.hse.gov.uk/msd/mac/. Last accessed 3 Nov 2020
8. NIOSH Lifting Calculator https://www.cdc.gov/niosh/topics/ergonomics/nlecalc.html. Last accessed 3 Nov 2020
9. Ergo/IBVErgonomics risk assessment software http://www.ergoibv.com/en/home/. Last accessed 3 Nov 2020
10. ErgoFellow 3.0 https://www.fbfsistemas.com/ergonomics.html. Last accessed 3 Nov 2020
11. Vitos Erzsébet (2019) Risk prevention of manual handling in the retail sector, thesis. Óbuda University, Budapest
12. Kovács Csaba (2020) Ergonomic development of three critical workplaces, thesis. Óbuda University, Budapest

New Proposals for Improvement of Train Driving Support Systems Performances

Hrvoje Haramina, Adrian Wagner, and Kristijan Dolenec-Čižmek

Abstract The modern railway traffic management practice increasingly implies application and development of train driving support systems. The focus of research in this area is on the impact of these systems on the efficiency of railway transport, mostly through the quality and optimization of railway timetables realization and energy efficient train driving. In this paper, an overview of the application of train driver support systems is given and their impact on the efficiency of railway traffic is analyzed. The results of the analysis showed some advantages of train driving support system application related to quality of timetable realization and energy efficient train driving but also marked possible problems related to increase of train driver workload and hence decrease of railway traffic safety. Based on this analysis, new train driving support systems functions for improvement of railway traffic efficiency are proposed. Additionally, a new proposal of systems interface design performance by using human voice for advice information transmission as an addition to displaying information on the monitor for the purpose of railway traffic safety improvement which can be reduced by increase of train driver workload and focus distraction due to application of train driving support system is suggested.

Keywords Train driver advisory systems · Driver-machine interface design · Ergonomics · Railway traffic efficiency

H. Haramina (✉) · K. Dolenec-Čižmek
Faculty of Transport and Traffic Sciences, University of Zagreb, 4 Vukelićeva, Zagreb, Croatia
e-mail: hharamina@fpz.unizg.hr

A. Wagner
Carl Ritter von Ghega Institute, St. Pölten University of Applied Sciences, Sankt Pölten, Austria

D. Sumpor et al. (eds.), *Proceedings of the 8th International Ergonomics Conference*, Advances in Intelligent Systems and Computing 1313,
https://doi.org/10.1007/978-3-030-66937-9_11

1 Introduction

The modern railway traffic management practice increasingly implies application and development of train driver support systems. Generally, train driving support systems i.e. train driver advisory systems (DAS) are on-board computer-based systems, and their purpose is to offer information of optimal driving strategy to the train driver. These systems do not represent safety systems but only give optional advices to the train driver how to optimally drive a train. Regarding to this, train driver should not accept recommendations offered by the system if he considers that in that case could endanger traffic safety. The main objectives of train driving support systems are in the first place positive impact on the traffic control process regarding better quality of timetable realization in terms of the high rate of railway line capacity utilization and as a secondary benefit an improvement of energy efficiency in train driving. Additionally, some other optimization criteria (e.g. pollution and noisy reduction in populated areas, better passenger comfort and satisfaction etc.) can be used for definition of optimal driving strategy [1]. The scientific research in this area is mostly focused on the development of new types of algorithms which can increase positive impact of such systems on efficiency criteria [2, 3].

Regarding the fact that modern railway traffic management practice increasingly implies application and development of these systems an analysis of their impact on railway traffic efficiency and safety should be conducted. Thus, in this research an existing practice of train driving support systems application will be presented and the influence of their performances on the railway traffic efficiency related to quality of timetable realization, energy efficiency and railway safety will be analyzed. Based on that, new driving support system functionalities as well as the design of associated interface characteristics will be proposed.

2 An Analysis of Train Driving Support Systems Application

Technically, there is a three basic types of train driving support systems: static, dynamic train-related and dynamic network—related type. Today, main focus is on the development of dynamic types of these systems.

Static systems can give just some offline pre-prepared information for a train driver about how to efficiently drive some certain train on the specified railway line, usually this consider information like when to start a coasting phase of driving if the train is on time i.e. it has no delay, which are based on the planned timetable data and not considering observed train delay or any kind of traffic disturbances. In the case when the train has delay the suggestion will be to continue its ride with shortest running time. In railway networks without possibility of continuous data transmission between trains and traffic control center dynamic train-related driver advisory systems can be an appropriate solution. This kind of support systems

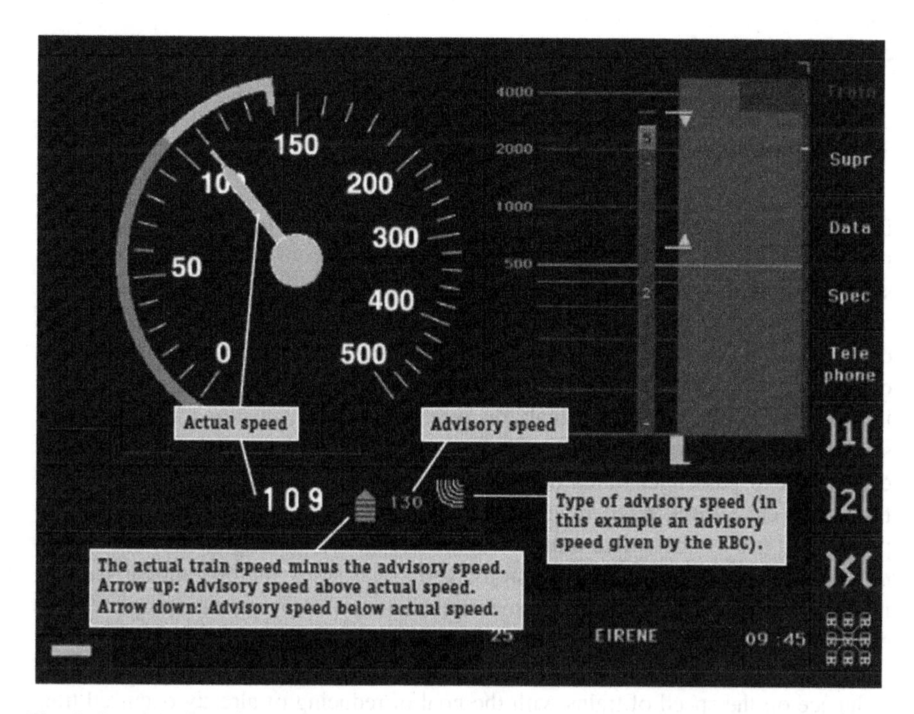

Fig. 1 Interface for train driving support information. *Source* Taken from TCS-MMI interface, http://www.humanefficiency.nl/etcs/etcs-dmi.php, [4]

considering observed train state but not information about the state of other trains in the network. In contrast, dynamic network-related systems are dedicated for the railway lines equipped with train control system based on continues data transmission between train and traffic control center e.g. ETCS (European Train Control System) level 2. In this case solutions offered by train driving support systems include overall traffic conditions regarding current speed and position of other trains in the network and sometimes even anticipations of their behavior in some near future. This is important for optimal use of infrastructure elements which could be used for train routes setting and giving movement authorities for observed train.

Train driver interface is usually performed with stand-alone display placed on driver desk or the information could be displayed on an ERTMS/ETCS standard display, with some minor modifications which can be additionally adjusted for this purpose (Fig. 1).

Due to the increasing need for improvement of rail transport efficiency with reducing operational costs a large number of different driver support systems is being developed e.g. CATO—Computer Aided Train Operation, Energy Efficient Timetabling at SNCF, Real-time rescheduling at SSB, FreightMiser, ESF EBuLa, Bombardier Driving Style Manager, Lötschberg base tunnel advisory speed system, Czech system AVV GEKKO, LEADER RouteLint, Trip Optimizer, System for Vienna Line U6 etc. [5].

3 Influence of Train Driving Support System Performances on the Railway Traffic Efficiency

3.1 Impact on Timetable Stability

Railway infrastructure managers are responsible for railway traffic management and timetable planning. Despite their efforts in construction of robust timetables with sufficient time reserves in train running times and headways, in the practice it is not always possible to avoid occurrence of train conflicts during rail operation. Such conflicts can be caused by various unpredictable traffic disturbances (e.g. extension of trains running time due to vehicle failures during the ride, infrastructure failures like problems with lineside signals or switches, fake occupations of some track section, failure of level crossing devices, problems with power supply system, occurrence of passenger trains delay due to extension of planned dwell times on train stops etc. Currently, there are different strategies for resolving this conflicts in praxis, e.g. use of alternative train routes, relocation of meeting stations, extension of scheduled dwell time on train stops, additional stops for operational requirements, extension of trains running time, cancellation of train over its complete route or some part of it.

Beside this, one of the possible strategies to overcome this problem is an influence on the speed of trains with the goal of reducing of already occurred train delay and disabling of the delay spreading on other trains in the network. For this purpose, train driving support systems can have very big positive influence on timetable stability. Thereby, application of train driving support systems along with quality distribution of regular time supplements for train delay recovery allows decrease in amount of buffer times while maintaining the same level of timetable realization quality which can result, especially in the case of train controlling by the use of cab signaling systems (e.g. ETCS level 2), in significant increase of railway network capacity. Thus, the appearance of new train conflicts can be avoided by anticipated influence on traffic control process by using algorithms based on artificial intelligence [6].

3.2 Impact on Energy Efficiency of Railway System

Modern methods for energy efficient train driving usually consider usage of computer-based driving support systems. These systems, based on train dynamics and track resistance data for specific railway line section, calculate optimal energy efficient speed profile i.e. running regime for specific train which needs to come to specific point of railway line within the running time selected by the traffic control system. This specific point is usually some stop sign for train head at the station or place in front of specified main signal which represent the end of given movement authority for this train.

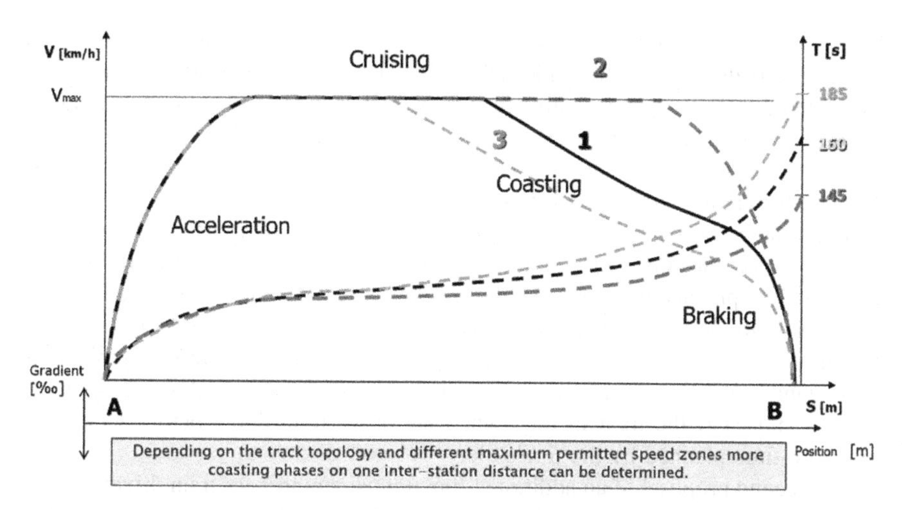

Fig. 2 Selection of the most appropriate train running regime

Usually, one typical speed profile for running of suburban train between two train stops will contain four different running phases (Fig. 2).

At the beginning of the trip train is usually running with maximal acceleration till it reaches its maximal permitted speed and then it is cruising with this speed. After cruising phase, train is coasting and finally braking till it stops.

In principle, energy efficiency can majorly be achieved by a more significant share of train coasting phase in its running regime. Thus, in energy efficient timetables initial running regimes of trains will contain a certain share of coasting which should be call off in the case when the train is late and has to run faster to catch up occurred delay as much it is possible.

3.3 *Proposal of New Support Functions and Interface Design for Improvement of Train Driving Support Systems*

The usage of train driving support systems could also have negative impact on railway traffic efficiency. Namely, previous research showed that usage of support systems in driving of certain train types can impose higher workload on drivers compared to a control (no DAS) condition [7, 8]. This can impact on railway traffic safety due to overload of cognitive workload of train drivers which can lead to safety critical mistakes in their actions during the train operation process. Based on this, suggestion is that support information of train driver support system, beside its visual presentation on some cab monitor interface, should be additionally presented as sound message performed by human voice. This kind of interface design can have positive impact on train safety because train driver can visually focus ahead the train to better observe line side safety elements like signals and signs and give more attention to the other control functions related to the tasks which should be performed by train driver located at the driver-desk. This kind of information presentation also can be applied for some additional train driver support functions, beside information about current train optimal speed.

The first proposed function is related to the quality of timetable realization. Namely, sometimes due to train driver's inattention train can depart earlier than it should by the planned timetable. To decrease possibility of this problem, additional information to the train driver can be given. This can be performed as information placed on the display used for application of driver support system in combination with sound interface by means of human voice which is giving advice to the train driver to not start the train in time when he closes train doors before scheduled departure time planned by the timetable.

The second proposal of an additional function is reminder for train driver to give sound sign when train is approaching to the level crossing which is not secured by level crossing safety device. This information can also be given by human voice interface. The third proposal is related to the case on the line with centralized traffic control when passenger train which already has movement authority till the exit signal starts from station platform and then must stop at the exit signal. In this situation train driving support system would give information to the train driver to start the train in the optimal time to avoid stopping at exit signal.

Previous research [9] has shown that train drivers have strong interest in the surrounding traffic, need up to date information about the traffic plan, and have valuable information that could improve operative planning which can improve their Situation Awareness (SA) of the current traffic situation. Based on this, our suggestion is that driver support system interface also can be used for sending information which can be useful for online timetable rescheduling process to the control center. For instance in the case of existing delay of e.g. suburban train or if the train driver assume that delay of his train will occur due to prolongation of dwell time on the current train stop, he can assume what will be the real departure time of his train from the stop. This assumption can be based on different factors, e.g. number of passengers in the train and at the platform, age and behavior of passengers he can see at the platform, weather conditions, eventually door closing problems etc. Namely, this kind of information can be valuable input data for timetable rescheduling process which is a background for work of train driving support systems in certain number of trains that could be directly or indirectly influenced by departure time of this suburban train.

In some cases, train drivers can purposely refuse to accept suggestions given by the driver advisory system, because they estimated that train can be operated more efficiently by application of different driving strategy than the one suggested by the system. This can be approved by the inability of the system to always offer better solutions than experienced train driver due to the structure of its algorithm which sometimes doesn't consider all possible inputs that can influence calculation of optimal solution which experienced and motivated train driver can recognize and consider in his decisions. On the other side, with better acceptance of suggested train driving information by train drivers the traffic management system can work with more reliable data necessary for anticipation of movement of all controlled trains in some short foreseeable future time. Because of this reason it is very valuable to increase train drivers experience in efficient train operation. This can be achieved by additional information for train drivers performed by driving support system which can improve train drivers understanding of current traffic situation and suggested driving strategy in the certain future travel period. For instance, this kind of information can be information about next goals which are planned to be achieved by application of suggested driving strategy.

4 Conclusion

In this paper an existing practice of train driving support systems application is given and the impact of its performances on the railway traffic efficiency is analyzed. The results of the analysis showed some advantages of train driving support system application related to quality of timetable realization and energy efficient train driving but also marked possible problems related to increase of train driver workload and hence decrease of railway traffic safety. Based on this analysis, new measures for improvement of functionality and interface design for increase of traffic safety are proposed. New proposals for improvement of systems functionality should have positive impact on railway traffic efficiency and voice based massaging as an additional interface between train driving support system and train driver can improve traffic safety which can be reduced by increase of train driver workload and focus distraction due to application of train driving support system.

References

1. Hansen IA, Pachl J (eds) (2014) Railway timetabling & operations. In Analysis—modelling—optimisation—simulation–performance evaluation, 2nd ed. Eurailpress
2. Lindner T, Milius B, Arenius M, Schwencke D, Grippenkoven J, Sträter O (2014) Considerations on the reliability of the driver: recording of safety-influencing factors and their significance on the basis of event data (Betrachtungen zur Zuverlässigkeit des Triebfahrzeugführers. sicherheitsbeeinflussender Faktoren und ihrer Bedeutung auf Basis von Ereignisdaten). EI-Eisenbahningenieur, Januar 12, pp 10–16
3. Yang L, Lidén T, Leander P (2013) Achieving energy-efficiency and on-time performance with driver advisory systems. In IEEE conference on intelligent rail transportation (ICIRT). IEEE, Beijing, pp 13–18
4. ETCS-MMI interface, http://www.humanefficiency.nl/etcs/etcs-dmi.php. Last accessed 19 Oct 2020
5. RSSB, https://www.rssb.co.uk/search#search-results_e=0&search-results_q=driveradvisory systems. Last accessed 19 Oct 2020

6. Haramina H, Mandić M, Nikšić M (2012) New method for energy-efficient train operation on commuter rail networks. Teh Vjesn Tech Gaz 19(4):801–806
7. Large DR, Golightly D, Taylor EL (2014) The effect of driver advisory systems on train driver workload and performance. In Proceedings of ergonomics and human factors conference (EHF2014), Southampton, UK, 7–10 April 2014
8. Haramina H, Ljubaj I, Toš I (2017) An analysis of passenger train driver's cognitive workload in relation to croatian national train control system (Analiza kognitivnog opterećenja strojovođe putničkog vlaka s obzirom na hrvatski nacionalni sustav vođenja vlakova). Sigurnost 59(2):99–108
9. Tschirner S, Andersson AW, Sandblad B (2013) Designing train driver advisory systems for situation awareness. In Dadashi et al (ed) Rail human factors: supporting reliability, safety and cost reduction. Taylor & Francis

3D Printing Technologies for Clothing Industry: A Review

Renata Hrženjak, Slavica Bogović, and Beti Rogina-Car

Abstract 3D printing represents a process for making a physical object from a three-dimensional digital model, typically by laying down many successive thin layers of a material. It brings a digital object into its physical form by adding layer by layer of materials. This technology is represented in a large number of different production sectors, especially in the modern aviation industry, construction industry, clothing industry, automotive industry, shipbuilding, medicine, dentistry, film industry and various design and artistic fields. The quality implementation of 3D printing technologies in each of the mentioned branches of production is important. The earliest 3D printing technologies first became visible in the late 1980s, and then they were called Rapid Prototyping (RP) technologies. This is because the processes were originally conceived as a fast and more cost-effective method for creating various prototypes for product development within industry. The paper presents the development of 3D printing technology as well as the opportunities it provides for the clothing and fashion industry.

Keywords 3D printing · Materials · Fashion and clothing industry

1 Introduction

3D printing is a relatively new manufacturing technology related to sustainability because it reduces waste [1]. 3D printing production is more environmentally friendly, so mastering this area is topical [2]. The starting point for any 3D printing process is a 3D digital model, which can be created using a variety of 3D software programmes—in industry this is 3D CAD, for Makers and Consumers there are simpler, more accessible programmes available—or scanned with a 3D scanner. The model is then 'sliced' into layers, thereby converting the design into a file readable by the 3D printer.

R. Hrženjak (✉) · S. Bogović · B. Rogina-Car
Faculty of Textile Technology, University of Zagreb, Zagreb, Croatia
e-mail: renata.hrzenjak@ttf.unizg.hr

D. Sumpor et al. (eds.), *Proceedings of the 8th International Ergonomics Conference*, Advances in Intelligent Systems and Computing 1313,
https://doi.org/10.1007/978-3-030-66937-9_12

The material processed by the 3D printer is then layered according to the design and the process. As stated, there are a number of different types of 3D printing technologies, which process different materials in different ways to create the final object. Functional plastics, metals, ceramics and sand are, now, all routinely used for industrial prototyping and production applications. Generally speaking though, at the entry level of the market, materials are much more limited. Plastic is currently the only widely used material—usually ABS or PLA, but there are a growing number of alternatives, including Nylon [3]. 3D technologies today occupy a significant place in the design of individualized objects. In recent years, intensive work has been done on the development and improvement of sophisticated equipment that enables 3D design of objects. This category includes 3D scanners that digitize existing objects or human bodies, computer software packages used to design 3D objects and 3D printers for making shaped objects [4–6]. For the purpose of designing 3D objects intended for 3D printing, software packages (Maya, Rhino3D, Solidworks, MeshLab, Blender, etc.) are used, which enable the construction and manipulation of 3D objects that can be stored in various data formats [4].

2 3D Printing Materials and Technology

3D printing of objects that are created, constructed and designed in software packages for 3D design can be made from different materials depending on the applied technique of 3D printing and the final purpose of the object. In 3D printing by the FDM/FFF (fused deposition modelling/fused filament fabrication) process, a molten polymer material is used which is layered. The most commonly used materials are: acrylonitrile/butadiene/styrene (ABS), polycarbonate (PC), polyamide (PA), thermoplastic polyurethane (TPU) and polylactic acid (PLA). When preparing a 3D model for 3D printing, it is possible to define the wall thickness of the object to be printed, as well as the method of filling and its density (Fig. 1). In this way, items of different strength and weight depending on the purpose can be made [4, 6, 7].

Selective laser sintering (SLS) is a rapid prototyping technology that uses powder to create models. The models created by this process are light and very precisely made. The SLS printing process begins by applying a laser to the initial layer of powder, which then hardens, and the other layers are connected to the previous one in the same way. The advantage of this technology is that it does not require the construction of a support during printing because the product is surrounded by powder at all times during production [8].

Fig. 1 Infill patterns at varying densities of the filling of the 3D object obtained by 3D printing [7]

2.1 Flexible Structure of Clothing Elements for 3D Printing

In order for the 3D printed object to have a flexible structure and the possibility of stretching in different directions (Fig. 2), in addition to the choice of material and method of 3D printing, it is necessary to select construction elements that will provide the necessary flexibility (Fig. 3) [9].

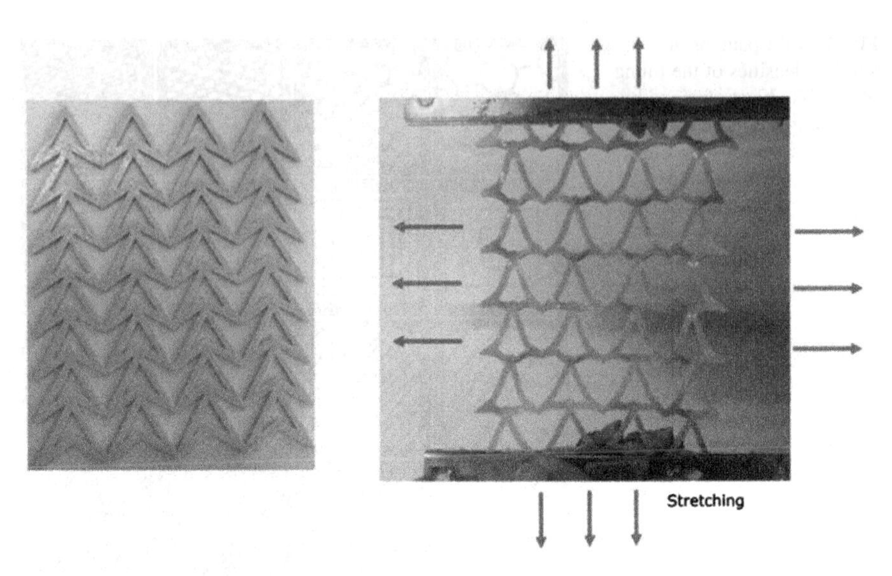

Fig. 2 Behavior of the 3D printed structure during stretching process along the x and y axes [9]

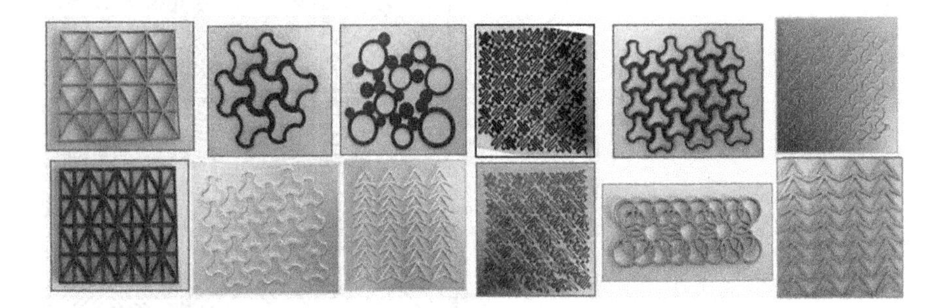

Fig. 3 Various geometries produced with different materials and colors [9]

3 3D Printing for Clothing and Fashion

Dutch designer, Iris van Herpen, is probably one of the most influential designers when it comes to mixing fashion and 3D printing. Many of her collections include 3D printed elements, from dresses to accessories. Thanks to 3D printing, designers have more freedom in terms of design and geometric complexity. In fact, they are able to produce shapes that are impossible to achieve using conventional production methods [10]. In 2013 Stratasys Ltd. (NASDAQ: SSYS), a leading manufacturer of 3D printers and production systems for prototyping and manufacturing and Materialise, a Belgian-based pioneer in Additive Manufacturing software and solutions, announced the unveiling of 3D printing collaborations on the catwalks of Paris Fashion Week as part of Iris van Herpen's Haute Couture show,

'VOLTAGE'. Van Herpen's eleven-piece collection featured two 3D printed ensembles, including an elaborate skirt and cape created in collaboration with artist, architect, designer and professor Neri Oxman from MIT's Media Lab, and 3D printed by Stratasys. An intricate dress was also designed in collaboration with Austrian architect Julia Koerner, lecturer at UCLA Los Angeles, and 3D printed by Materialise, marking the second piece created together with Koerner and the ninth with Materialise (Fig. 4a, b) [11].

Iris van Herpen also uses PolyJet 3D printing technology directly onto textiles—representing an interesting breakthrough for the industry. By combining traditional textile materials with digitally-created 3D printing materials, she's hoping to bridge the gap and facilitate a faster integration of the technology in textile design. Figure 4c presents foliage dress, part of the Ludi Naturae haute couture collection from Van Herpen, developed by TU Delft scientists and 3D printed using Stratasys PolyJet technology. Three variations of the 3D printed material were altered on a droplet level, achieving the unique color and transparency on the dress which allows it to seamlessly fuse with the fabric material. Another benefit of integrating 3D printing and textiles together, is that it enhances the practicality and comfort of the garment for the wearer. The interface that touches the skin of the wearer can be the soft fabric, while the complex 3D printed design elements can be enjoyed on the outer part of the garment—enhancing the comfort for the wearer and addressing this previous drawback [12]. Figure 4d presents 3D design combined with traditional handcraft. Collaboration was organized between Iris van Herpen and Benthem Crouwel Architects. Benthem Crouwel's design for a new extension to Amsterdam's Stedelijk Museum had earned it the nickname "bathtub". This inspired van Herpen to design a dress that would fall around the wearer like a splash of water, like being immersed in a warm bath [13].

Israeli designer Danit Peleg (recognized by Forbes as one of Europe's Top 50 Women in Tech) [14] and her team are eager to keep pushing the boundaries of 3D

Fig. 4 3D printing of clothes by Iris Van Herpen [11–13]

Fig. 5 **a** Design and 3D print fashion based on 3 different approaches, **b** collection "Liberty leading the people"

printed fashion. They are passionate about challenging the ecosystem to develop new materials, new printing techniques, and new software breakthroughs (Fig. 5a, b). In the summer of 2017, Danit launched a customization and personalization platform on her website, enabling customers to personalize and order the world's first 3D printed garment available to purchase online. This venture has led to Danit's latest undertaking, in teaching a virtual workshop for beginners on 3D Printing Fashion—the very first of its kind. In the future, as the technology improves, Danit envisions a world where anyone will be able to buy files and print clothes at home or at designated stores [15, 16].

4 Conclusion

People sometimes mention the notion of 3D printing reaching the mass-market, but it is known that caution must be taken when considering the implications of mass-market solutions. Ultimately, it is hoping that the fashion world returns to a more sustainable model, which involves more localized production, allowing smaller design and production houses to compete in the market. In a world of constant technological advancements, fashion can also be considered as a key vehicle for demonstrating the vast capabilities of 3D printing for design in other sectors—whether it be consumer goods, automotive, aerospace or others. Not everybody can connect with technology when it derives from a very particular niche market. However, fashion opens up a new way of relating to technology and allows greater participation [17]. Although 3D printed fashion still is a thing of luxury the technology has great potential to expand into the realms of mass-production and customization. Some companies are already offering 3D printed products through online shops and configurators, opening the door to many more possibilities in the future. And with new innovative techniques designers are learning to overcome the material rigidity inherent to 3D printing. It will soon be possible to buy 3D printed products in local clothing stores [18]. But despite all the advantages, 3D technology also has some challenges that need to be taken into account. For many manufacturers and end users it is difficult stating with certainty that parts or products produced by 3D printing will be of consistent quality, strength and reliability because it isn't standardized. Additive technology itself very much impacts the environment and also, required equipment and product costs are high. These are all challenges that need to be overcome in the future.

References

1. Gebler M, Schoot Uiterkamp AJM, Visser C (2014) A global sustainability perspective on 3D printing technologies. Energy Policy 74:158–167
2. The effect of 3D printing on a textile fabric. https://www.researchgate.net/publication/341558138. Last accessed 02 Sept 2020
3. https://3dprintingindustry.com/3d-printing-basics-free-beginners-guide#03-technology. Last accessed 02 Sept 2020
4. Bogović S (2018) New technologies for designing protective clothing and equipment. In: Kirin S (ed) 7th international professional and scientific conference. Karlovac University of Applied Sciences, Zadar, Croatia, pp 578–584
5. Maurer M Vitus 3D body scanner. http://www.3dbodyscanning.org/cap/papers/A2012/a12009_08maurer.pdf. Last accessed 02 April 2020
6. Godec D et al (2018) Utjecaj parametara 3D tiskanja na savojna svojstva 3D tiskanog proizvoda, Zbornik radova 11. In Ercegović Ražić S, Martinia Ira Glogar M, Novak I (eds) Znanstveno–stručnog savjetovanja Tekstilna znanost i gospodarstvo. Sveučilište u Zagrebu Tekstilno-tehnološki fakultet, Zagreb, Hrvatska, pp. 74–79
7. Slic3r Manual: https://manual.slic3r.org. Last accessed 10 Mar 2020
8. Ružić A, Novak L (2020) 3D printeri. https://anaruzic.wixsite.com/3dprinteri/vrste. Last accessed 18 Sept 2020
9. Spahiu T, Canaj E, Shehi E (2020) 3D printing for clothing production. J Eng Fibers Fabr. https://doi.org/10.1177/1558925020948216. Last accessed 15 Sept 2020
10. https://www.3dnatives.com/en/iris-van-herpen-110220195/. Last accessed 05 Oct 2020
11. https://www.materialise.com/en/cases/iris-van-herpen-debuts-wearable-3d-printed-pieces-at-paris-fashion-week. Last accessed 05 Oct 2020
12. https://www.whichplm.com/rise-3d-printing-fashion/. Last accessed 06 Sept 2020
13. https://www.irisvanherpen.com/haute-couture/crystallization. Last accessed 05 Oct 2020
14. https://www.forbes.com/profile/danit-peleg/#375aa9f1ecd8. Last accessed 08 Oct 2020
15. https://danitpeleg.com/about/. Last accessed 26 Sept 2020
16. 3D Printing for garments production: an exploratory study. https://www.researchgate.net/publication/312309878. Last accessed 15 Sept 2020
17. Mageean L. The Rise of 3D Printing in Fashion. https://www.whichplm.com/rise-3d-printing-fashion. Last accessed 09 Oct 2020
18. https://all3dp.com/2/3d-printed-fashion-the-state-of-the-art-in-2019. Last accessed 10 Sept 2020

Research of Working Postures in the Technological Sewing Process Using the REBA Method

Snježana Kirin and Anica Hursa Šajatović

Abstract The characteristics of the work in the technological sewing process depend on the type and character of the workpiece, the type of the sewing machine used to perform the technological operation, as well as on the training and skills of the worker. In the technological sewing process, the work is performed in a sitting position at the sewing machine, whereby the worker performs hand and/or machine-hand technological suboperations. During the performing of technological operations, the worker uses her hands and torso to handle the workpiece, and the feet to operate the sewing machine pedal at short intervals with continuous repetition throughout the work shift, subjecting the worker to stress. The paper analyses the working position at nine workplaces in the technological sewing process using the REBA method (Rapid Entire Body Assessment). The analysis of the results showed that there is workload (medium or high risk) at all workplaces that requires workplace adaptation in the form of workplace redesign. The redesign of workplaces involves adapting the dimensions of the workplace to the worker's height and determining a more favourable working method that achieves a lower workload.

Keywords REBA method · Technological process of sewing · Worker workload

1 Introduction

In the sewing process, technological operations are performed on installed production lines with the existing arrangement of machines and devices depending on the type of garment [1]. The work is performed in a sitting position at the sewing

S. Kirin (✉)
Karlovac University of Applied Sciences, Karlovac, Croatia
e-mail: snjezana.kirin@vuka.hr

A. H. Šajatović
Faculty of Textile Technology, University of Zagreb, Zagreb, Croatia
e-mail: anica.hursa@ttf.unizg.hr

D. Sumpor et al. (eds.), *Proceedings of the 8th International Ergonomics Conference*, Advances in Intelligent Systems and Computing 1313,
https://doi.org/10.1007/978-3-030-66937-9_13

machine, where the worker uses the body and hands to handle the workpiece, and the legs to control the pedal of the sewing machine. With such a system there is a risk of damage to the bone and joint system due to loading positions and/or movements (bent or curved trunk and head, bent wrists) carried out at short intervals with continuous repetition during the work shift. The risk of injuries and damages to the bone and joint system increases due to the lack of training of workers and with increasing age and the number of years of working in a sitting position. It most frequently manifests itself in the form of back and neck pain, changes in the joints, inflammation and damage to tendons, muscles and nerves in the arms [2]. In daily practice, it is therefore necessary to monitor work processes and try to find better solutions taking into account all the relationships in the man-machine-environment system [3]. Preventive measures to reduce the workload of workers and to prevent the occurrence of bone and joint disorders are the ergonomic design of the workplace, work using an appropriate work method and education of workers.

Workplace design and work methods result in dimensional harmony of the man-machine-environment system. It features a correct seating position that allows fast and precise engine movements when starting the machine and guiding the workpiece, a high degree of coordination, a correct lumbar, back and neck position and a good head position [4–6].

The aim of this paper is to show the workload of workers in the real production process of making men's jacket using the REBA method.

2 REBA (Rapid Entire Body Assessment) Method

Hignett and McAtamney (2000) developed the REBA (Rapid Entire Body Assessment) method, whereby a matrix of 144 different body postures is used in order to determine unfavourable working postures during the work that are summed up after observation and the level of risk or workload at the workplace is determined [7]. The computer system ErgoFellow was developed, which contains a module of the REBA method according to the specifications of the authors Hignett and McAtmney [8].

To assess the body posture (neck, back, legs), there are three neck positions (+ additional position for forward bending and rotation), five back positions (+ additional position for forward bending and rotation) and two leg positions (+ additional two positions for lower leg angle). In addition, it contains three levels of body load (additional position of strength). In the case of the arm positions, there are five upper arm positions (+ three positions in relation to adjusting), two lower arm positions, two hand positions (+ one rotation position) and four options of grasping movement. Furthermore, there are three possibilities of activity and dynamics of the work task. Based on the analysis of the specific posture per body parts, the ErgoFellow computer system module of REBA method provides a summary assessment and degree of risk at the workplace (Fig. 1).

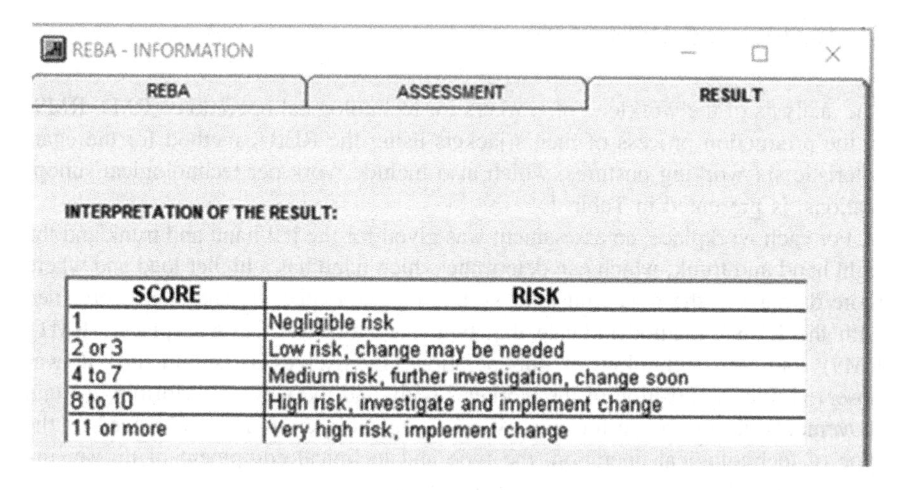
REBA - INFORMATION

REBA | ASSESSMENT | RESULT

INTERPRETATION OF THE RESULT:

SCORE	RISK
1	Negligible risk
2 or 3	Low risk, change may be needed
4 to 7	Medium risk, further investigation, change soon
8 to 10	High risk, investigate and implement change
11 or more	Very high risk, implement change

Fig. 1 Presentation and interpretation of the results (risk level) of the score of body load according to the REBA method [8]

3 Experimental Part

In order to study the workload of workers in the real production process, nine workplaces (RM1–RM9) within the production line of manufacturing men's jackets in clothing manufacturing company Varteks, Varaždin were analysed. At the analysed workplaces the following technological operations were recorded:

1. Sewing the back seam of the sleeve lining
2. Sewing the lining on the underlay
3. Sewing the lining over the sleeve length
4. Sewing the back seam
5. Sewing the front and back part
6. Sewing the front and side part
7. Sewing the shoulder seam
8. Sewing the back seam of the sleeve
9. Closing the sleeve of the fabric and lining with leaving the opening.

The recordings were taken with an EOS 750D camera with a built-in 18–135 mm EFS video-capable lens. The workplaces were recorded for 20 min, which resulted in an average of between 15 and 25 performances for each technological operation. The workers were recorded in the sagittal plane on the left or right side, depending on the type of technological operation and the way the workplace was designed. The workload analysis (RM1–RM9) was carried out using the ErgoFellow computer system (REBA method module) after technological subop-erations: taking, putting together, positioning, sewing, seam alignment during the sewing break, laying off, and the total level of workload and risk for each workplace was determined.

4 Results and Discussion

The analysis of the workload of workers for technological operations (RM1–RM9) in the production process of men's jackets using the REBA method for the characteristic six working postures, which also include work per technological suboperations, is presented in Table 1.

For each workplace, an assessment was given for the left hand and trunk and the right hand and trunk, which can determine which hand has a higher load and where more damage to the bone and joint system can be expected. The results obtained with the REBA method indicate that both the workload per workplaces (RM1–RM9) and per technological suboperations themselves is significant. It was observed that the position of the worker's body and the way of performing certain movements during the performing of technological operations is influenced by the type of technological operation, the type and technical equipment of the sewing machine and the way of designing workplaces.

The load in the technological sub-operation of taking for the left hand (Fig. 2) received a grade of 4 (11.1%) for one workplace, 5 (33.3%) for three workplaces, 6 (33.3%) for three workplaces and 7 (22.3%) for two workplaces, which is a medium risk area and will soon require a job redesign. Loading of the right hand got a score of 3 (33.3%) for three workplaces resulting in a low risk, and a score of 5 (22.3%), 6 (33.3%) and 7 (11.1%) meaning a medium risk at six workplaces and requires a redesign of the workplace soon. In this kind of suboperation the worker uses the left hand in the maximum work zone to reach the workpiece resulting in a higher hand load, while with the right hand she grasps the workpiece in the central work zone.

In the technological suboperation of **putting together** the worker uses both hands to join two parts with an accuracy of ±1 mm, often with an increased front flexion of the spine and head. Precisely because of the high accuracy required for putting together two workpieces, the high focus of the view and the use of finger

Table 1 Presentation of the scores for work load per individual suboperations for the investigated workplaces

Technological suboperation	Score of workload																	
	RM1		RM2		RM3		RM4		RM5		RM6		RM7		RM8		RM9	
	L	R	L	R	L	R	L	R	L	R	L	R	L	R	L	R	L	R
Taking	4	3	5	3	5	5	6	6	6	5	5	3	7	6	6	6	7	7
Putting together	7	7	8	7	5	5	8	8	7	7	5	5	8	8	7	7	5	5
Positioning	7	7	8	7	5	5	8	8	7	7	8	8	8	7	7	8	7	7
Sewing	7	7	8	8	7	7	8	8	7	7	8	8	8	7	7	8	7	7
Alignment during sewing break	5	5	5	5	5	5	7	7	7	7	5	5	4	4	5	5	7	7
Laying off	5	6	7	6	5	6	3	3	5	5	5	4	8	8	7	8	6	6

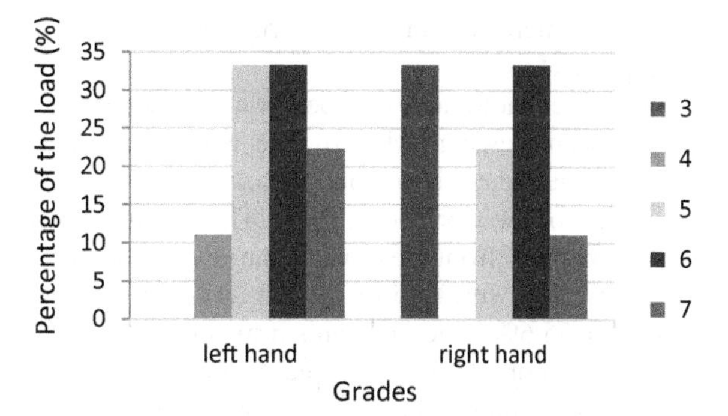

Fig. 2 Presentation of the percentage for the suboperation of taking

and hand movements of both hands, the load scores are higher (from 5 to 8). For the left hand, the workload got a score of 5 (33.3%) for three workplaces and 7 (33.3%) for three workplaces, which presents a medium risk and requires a redesign of the workplace soon, while three workplaces got a score of 8 (33.3%), which presents a high risk and requires an immediate redesign of the workplace. The right hand workload got a score of 5 (33.3%) for three workplaces and 7 (44.4%) for four workplaces, which presents a medium risk and requires a redesign of the workplace soon, while the workload for two workplaces got a score of 8 (22.3%), which requires an immediate redesign of the workplace (Fig. 3).

In the technological suboperation of **positioning** the worker uses both hands to position the workpiece under the sewing needle in the front sitting position with an increased flexion of the spine and the head with frequent unfavourable hand positions. The workload of the left hand got a score of 5 (11.1%) for one workplace, and four workplaces got a score of 7 (44.4%) which presents a medium risk and requires a redesign of the workplace soon. Four workplaces got a score of 8 (44.4%) which presents a higher risk and requires an immediate redesign of the workplace. The workload of the right hand got a score of 5 (11.1%) for one workplace and a score of 7 (55.6%) for five workplaces which presents a medium risk and a redesign of the workplace soon, while three workplaces got a score of 8

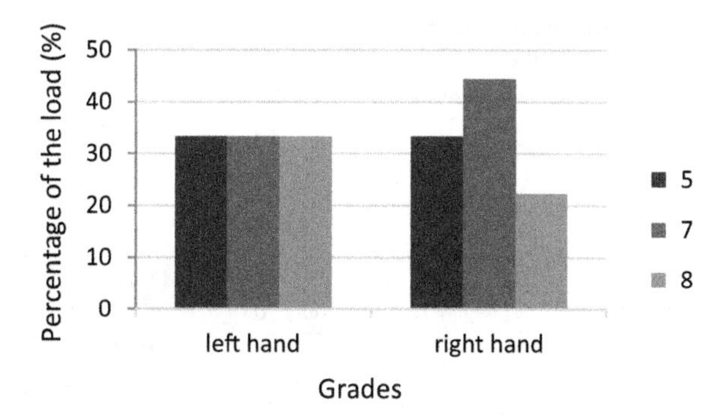

Fig. 3 Presentation of the percentage for the suboperation of putting together

(33.3%) which presents a high risk and requires an immediate redesign of the workplace (Fig. 4).

In the technological suboperation of **sewing** the worker guides the workpiece during seam sewing, whereby the worker monitors the seam position from the edges of the material, the mutual position of the material edges and the position of the length of the workpiece layers. These requires work in the front sitting position with enhanced front flexion of the spine and the head and unfavourable hand positions. The workload of the left and right hand got a score of 7 for five workplaces or 55.6% which presents a medium risk and it is necessary to redesign the workplace soon. At four workplaces the workload of the worker got a score of 8 or 44.4% which presents a high risk and an immediate redesign of the workplace is necessary (Fig. 5).

In the technological **suboperation during sewing break** the worker performs the alignment of the workpiece with both hands whereby sits in the front or medium sitting position depending on the technological operation and the demanding level of the suboperation of alignment. The workload of the left and right hand got a score of 4 (11.1%) for one workplace, a score of 5 (55.6%) for five workplaces and a score of 7 (33.3%) for three workplaces which presents a medium risk and it is necessary to redesign the workplace soon (Fig. 6).

In the technological suboperation of **laying off** at all workplaces the workers lay off workpieces on the movable stand on the right side, and mostly sitting in the back sitting position, but moving the trunk and the head. The workload of the left and right hand got a score of 3 (11.1%) at one workplace which presents a low risk. The workload of the left hand got a score of 5 (44.4%) at four workplaces, a score of 6 (11.1%) at one workplace and a score of 7 (22.3%) at two workplaces which presents a medium risk and requires a redesign of the workplace soon. The workload of the worker at one workplace for the left hand got a score of 8 (11.1%) which presents a high risk and requires an immediate redesign of the workplace. The workload of the right hand at one workplace got a score of 4 (11.1%), a score of 5 (11.1%), a score of 6 (44.4%) at four workplaces which presents a medium risk and requires a redesign of the workplace soon. The workload of the right hand at

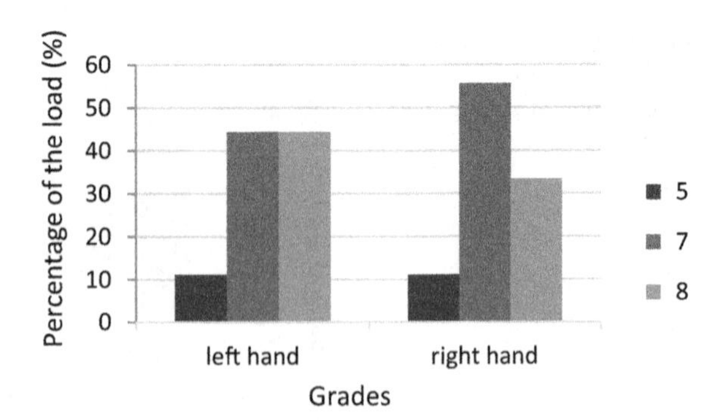

Fig. 4 Presentation of the percentage for the suboperation of positioning

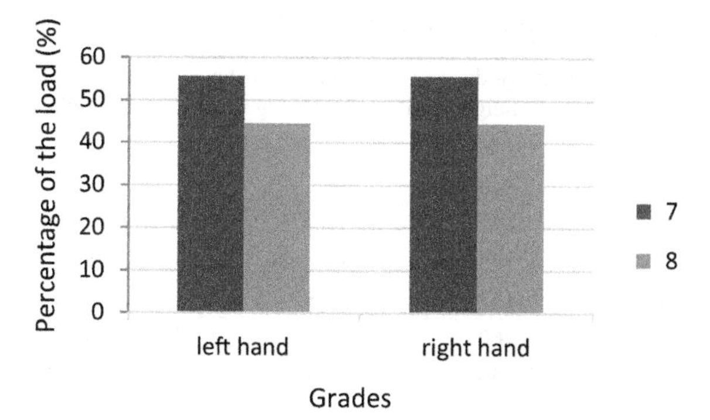

Fig. 5 Presentation of the percentage for the suboperation of sewing

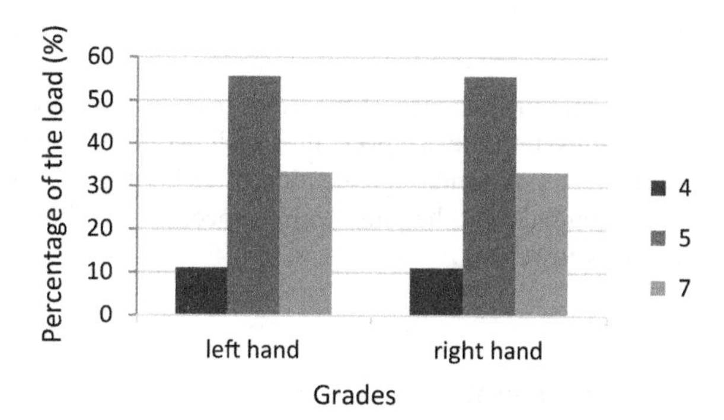

Fig. 6 Presentation of the percentage for the suboperation during sewing break

two workplaces got a score of 8 (22.3%) at two workplaces which presents a high risk and requires an immediate redesign of the workplace (Fig. 7).

In order to reduce the workload of workers in the technological sewing process per workplaces (RM1–RM9), it is necessary to adapt the dimensions of the workplace to the height of the workers. The height of the industrial seat results from the popliteal length of the lower leg with an addition of shoe thickness (2 cm) and pedal height (5 cm). The height of the working surface of the machine should be higher 10–15 cm than the thickness of the thigh in the sitting position of the

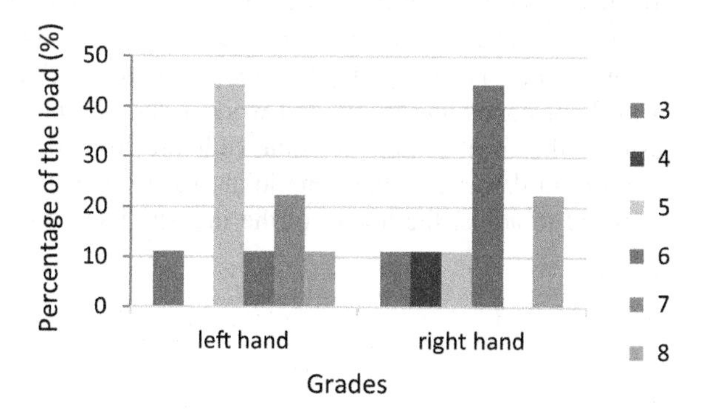

Fig. 7 Presentation of the percentage of the scores for the suboperation of laying off

Table 2 Recommended height of the work seat compared to the workers' body height

Workplace	Body height (cm)	Length of the lower leg (cm)	Actual seat height (cm)	Recommended seat height (cm)
RM1	160	42	51	49
RM2	160	42	54	49
RM3	162	42.5	51	49.5
RM4	178	46	56	53
RM5	167	43.5	53	50.5
RM6	160	42	54	49
RM7	156	41	51	48
RM8	167	43.5	52	50.5
RM9	163	42.5	53	49.5

worker, and the distance of the trunk from the working surface is 15–20 cm. These dimensions provide a space in which workers have the freedom of movement necessary for the smooth execution of technological operations. Table 2 shows the recommended seat heights with respect to the body height and length of the worker's lower leg.

5 Conclusion

The analysis of the results collected with the REBA method at nine workplaces in the production of men's jackets revealed that the load on the trunk, neck and hands were exist at all workplaces. Most workload scores are between 5 and 7, meaning a medium risk. In the technological suboperations: putting together, positioning and sewing, which are performed in the front sitting position, front flexion of the trunk, high concentration of view and high accuracy of the performing of the suboperation are necessary. Therefore, in these suboperations the workload generally gets a score of 8, which presents a high risk. In the technological suboperation of taking the right hand comes in low risk area (score 3) at three workplaces, and in the technological suboperation of laying off the left and right hand comes in low risk area (score 3) at one workplace. The characteristic of the work in the technological sewing process is a large number of repetitions of technological operations resulting in loads of hands and trunk, and work in the non-physiological position of sitting (usually the front sitting position with the front flexion of the head and trunk) resulting in damage to the bone-joint system. For the mentioned workplaces, it is proposed to adjust the height of the industrial seat to the worker's body height.

References

1. Šaravanja B, Hursa Šajatović A, Dragčević Z (2018) Istraživanje uvjeta radne okoline u tehnološkom procesima proizvodnje odjeće. Tekstil 66(5–6):146–154
2. HZZZSR. Praktična smjernica za ocjenu rizika pri ručnom rukovanju teretom. http://www.hzzzsr.hr/wpcontent/uploads/2016/11/Prakticna_smjernica_za_ocjenu_rizika_pri_rucnom_rukovanju_teretom.pdf. Accessed 2020/06/01
3. Shah ZA et al. Ergonomic risk factors for workers in garments manufacturing—a case study from Pakistan. http://ieomsociety.org/ieom_2016/pdfs/291.pdf. Accessed 2020/05/22
4. Jurčević Luić T, Runjak M (2013) Procjena opterećenja radnika pri podizanju tereta. Sigurnost 55(2):125–131
5. Kirin S, Dragčević Z, Firšt Rogale S (2014) Workplace redesign in the computer-aided technological sewing process. Tekstil 63(1–2):14–26
6. Kirin S, Hursa Šajatović A (2018) Defining standard sets of motions in the technological suboperation of sewing using the MTM-system. Tekstil 67(7–8):201–217
7. Hignett S, McAtamney L (2000) Rapid Entire Body Assessment (REBA). Appl Ergon 31(2): 201–205
8. Software Ergonomics—FBF Sistemas. https://www.fbfsistemas.com/ergonomics.html. Accessed 2020/05/01

Analysis of Working Conditions and Modeling of Activity Algorithms for Contact-Center Operators

Evgeniy Lavrov, P. Paderno, O. Siryk, E. Burkov, S. Kosianchuk, N. Bondarenko, and V. Kyzenko

Abstract The article examines the issues of ensuring the reliability of the activities of information processing operators who work in contact centers. A study of factors that reduce the effectiveness of call center operators was carried out. A method of formalized description of operators' activities based on functional networks of professor A. Gubinsky is proposed. The use of this method makes it possible to assess reliability taking into account the structure of the operators' activities and working conditions. A model is proposed to justify the number of operators, design the structure of activities and working conditions. The study can be useful to justify the feasibility of improving ergonomics in contact centers of commercial and government organizations.

Keywords Contact-center · Reliability · Simulation · Ergonomics · Working conditions · Man-Machine system

1 Introduction

For more than half a century of the existence of call centers and contact centers, a developed market for mass acceptance of telephone and Internet calls (both calls from customers in the service sector and requests from control systems of various levels) has formed [1, 2]. This is a globalized market: many Western companies use

E. Lavrov (✉)
Sumy State University, Sumy, Ukraine
e-mail: prof_lavrov@mail.ru

P. Paderno · E. Burkov
St. Petersburg Electrotechnical University «LETI», St. Petersburg, Russia

O. Siryk
Taras Shevchenko National University, Kyiv, Ukraine

S. Kosianchuk · N. Bondarenko · V. Kyzenko
Institute of Pedagogy of the National Academy of Educational Sciences of Ukraine, Kyiv, Ukraine

D. Sumpor et al. (eds.), *Proceedings of the 8th International Ergonomics Conference*, Advances in Intelligent Systems and Computing 1313,
https://doi.org/10.1007/978-3-030-66937-9_14

the services of contact centers located, for example, in India or Malaysia. Despite the seeming ease of work, due to the absence of heavy physical labor, the operators of such centers work under conditions of time pressure, tension and even stress [3, 4]. And the emergence of COVID-19 caused not only a surge in this area of business, but also the aggravation of a number of inherent problems.

2 Problem Statement

Despite the huge amount of research, for example [1, 2, 5–7], devoted to improving the efficiency of contact centers (the list of topics is extensive: automation, decision support systems, the use of robots and answering machines, office equipment and room design, relaxation technologies and etc.), the tasks of ergonomic support for the activities of such centers [7, 8] require new formally substantiated methods focused on quantitative assessments. These methods should enable prompt decision-making to ensure acceptable working conditions and, as a result, guarantee high efficiency in performing the functions required from contact centers.

Based on this, the problem statement can be formulated as follows—on the basis of the fundamental principles of ergonomics [8–11]: (1) identify the problems of providing working conditions for operators of contact centers; (2) develop principles for constructing descriptive models of the activity of contact center operators, necessary to predict the quality of activities and substantiate measures aimed at improving working conditions.

3 Methods

The proposed methodology for ergonomic support of contact center operators is based on an anthropocentric approach focused on the priority of the man-operator in the «human-technology-environment» system [12–16], with the obligatory construction of models that ensure the forecast of error-free and timely execution of requests and taking into account the following factors:

- working conditions;
- characteristics of hardware, software and interfaces;
- characteristics of operators (preparedness, functional state, motivation, etc.);
- organizational and technical characteristics (number of operators, activity settings, methods of distributing requests and organizing queues);
- technology of activity (including self-control);
- organization of quality control of operators-executors by operators-managers.

Despite the fact that there is a special system of standards and ergonomic norms for organizing the activities of operators, business owners often do not want to comply with all its requirements, since this is associated with significant costs. The proposed methodology provides for the formation of an explanatory component that makes it possible to substantiate the positive impact of the proposed ergonomic measures on economic efficiency.

4 Results

4.1 Global Challenges Negatively Affecting the Effectiveness of Contact Centers and Analysis of Working Conditions at Operator Workplaces

A survey of operators was conducted (with the participation of student Shapochka Yulia and graduate student Krivodub Anna), consisting of two parts:

- Global challenges that negatively affect the effectiveness of contact centers (in the context of Covid-19). Period of the survey: June–September 2020, number of respondents: 297 respondents from Ukraine and Russia. The results are shown in Table 1.

Table 1 Assessments of the significance of factors that negatively affect the effectiveness of contact centers (results of processing operator opinions)

Problematic issues of organizing the activities of operators of contact centers	Percentage of operators who consider the problem urgent
Insufficient automation of the processes of distribution and processing of applications (including decision support)	40.1
Lack of a single window for processing contacts through various channels	36.4
Usability issues and difficult to use databases (CRM)	29.9
Lack of resources (technical and human) to service all incoming requests	27.3
Lack of proper training (coaching) of operators	26.8
Lack of customer confidence in the system when communicating with robots (creates an additional burden on operators)	26.2
The need to be distracted by secondary work	19.8
The complexity of organizing remote (home) work (requirement due to the impact of COVID-19)	19.1
Poor (inadequate and (or) non-ergonomic) equipment of workplaces	13.5
Lack of feedback and the difficulty of promptly obtaining information about customers	10.1

- The negative (according to the operators themselves) influence of the factors of the working environment. Period of the survey: June 2015–September 2020, number of respondents: 517 respondents from Ukraine and Russia. The online questionnaire (in Russian) used in the survey is located at https://goo.gl/0j2ulZ. The results are shown in Fig. 1.

Obviously, in different contact centers, depending on national and professional specifics, the priority of problem areas and negative factors will be different. Therefore, separate studies are needed to eliminate the existing negative impacts on man-operators and, as a result, on the effectiveness of the contact centers themselves. The study showed that where the required effective feedback is established and the wishes of the operators are taken into account (67.3% of the contact centers surveyed by us), the operators' assessments of the attractiveness of their working conditions increase significantly within a short time (2–3 months).

For the integral scoring (0–60 points) and determination of the category of labor severity (1–6), as well as certification of workplaces, the authors have developed a special computer program [17].

4.2 *Principles of Modeling the Process of Fulfillment of Requests by Contact Center Operators*

In addition to the above factors, the structure of the operators' activities also significantly affects the reliability of the execution of requests by the operators of

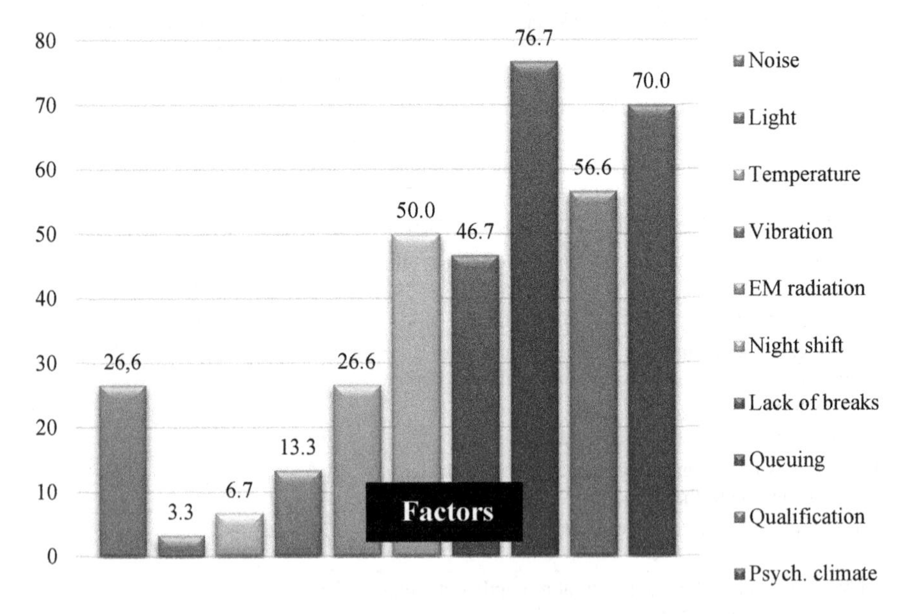

Fig. 1 The results of assessing the degree of negative impact on operators of physical and psychological factors of the workplace (according to operators' estimates)

contact centers. The organization of the execution of requests can usually be represented in the form of the activity algorithm [18].

The most convenient apparatus for describing and assessing the reliability of activity algorithms are functional networks [18, 19] on the basis of which the authors have developed the methodology for describing and evaluating the activities of contact center operators.

The value of the proposed model for describing activities is that it allows to obtain estimates of the reliability and time characteristics of all activities for execution of requests. To do this, it is required to use the software developed by the authors [19], having previously generated for each operation data on the probability of its error-free execution, as well as the mathematical expectation and variance of its execution time using either reference data or statistical databases of the enterprise (taking into account the values of all factors affecting working conditions, characteristics of operators, technical equipment, etc.).

The developed examples of assessing the reliability of the operators of contact centers for controlling access to distributed information resources are given in [20]. It also shows how working conditions affect the final results of activities. An example of such estimation is shown in Fig. 2.

In the case of a random flow of requests, operators are considered as specific ergatic «servicing devices», which are proposed to be modeled using functional networks, but also to use models of request queues, the lengths of which are determined, among other things, by the number of operators, their qualifications and other characteristics.

Thus, the developed computer program [21] is built according to the principle «Queuing system (simulation model of the processes of arrival, waiting in queues and fulfillment of requests) + functional network (model of the algorithm of activity for fulfilling the request)» and allows, by varying the organizational characteristics of the system, to choose the optimal variant of the organization that provides the specified conditions (according to the criterion of error and the criterion of timeliness, taking into account both the execution time and the waiting time in the queue) while fulfilling the ergonomic requirements (the coefficient of the queue of applications for the operator, the average length of the queue for the operator, the coefficient of employment of the operator, etc.).

The use of these models makes it possible to determine:

- required number of operators;
- how operators should fulfill requests (setting for error-freeness or performance);
- working conditions of operators;
- how to distribute requests between operators.

As a result, the task of justifying the costs of ensuring the effectiveness of the contact center can be solved (for this, a computer system was developed [22]).

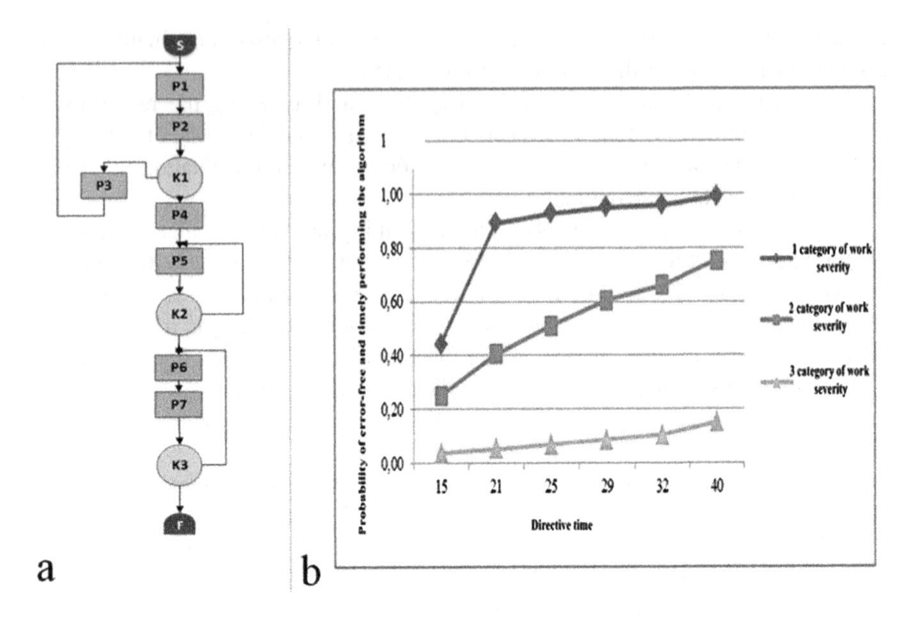

Fig. 2 An example of computer modeling using information technology: **a** functional network that simulates the process of executing an application (designations—according to [16], rectangle—working operation, circle—control of functioning); **b** dependence between the directive execution time of the application and the probability of timely and error-free execution of the algorithm (for various categories of work severity [16, 17]), modeling was carried out by Anna Krivodub

5 Conclusions

The widespread use of contact centers has made the problem of their ergonomic support urgent.

The considered task of studying the assessment by operators of working conditions and predicting the reliability of the execution of requests is solved by the complex use of methods of expert assessment and modeling of the activities of operators based on the activity approach.

For the effective functioning of contact centers, the following should be organized:

- feedback on the study of operators' assessments of working conditions;
- assessment of the influence of a set of factors (working conditions, structure of activities, number and qualifications of operators, characteristics of the flow of applications, etc.) on the ergonomic characteristics of the system and the effectiveness of the contact center.

The scientific novelty of the result lies in the fact that, in contrast to the available (usually intuitive and empirical) methods of ergonomic support, the proposed approach is based on a set of analytical and simulation models built according to the methodology of functional networks.

The practical result is the technology of studying the assessments of their working conditions received from the operators, as well as the computer modeling technology, built on the principle of «*what will happen if*» .

The reliability is confirmed by a wide approbation of the approach in the design and operation of polyergatic control systems (banking systems, software management systems, utility payment centers, systems of access to information resources).

Acknowledgements The authors dedicate this article to the memory of their teacher the first president of the Soviet Ergonomic Association, Doctor of Technical Sciences, Professor Anatoly Ilyich Gubinsky (1931–1990, St. Petersburg, Russia), who was the founder of the scientific school "Efficiency, quality and reliability of man-machine systems" and who first formulated the ideas that formed the basis of our study.

References

1. Sherstneva A, Sherstneva O (2018) The generalized model development for call center functioning. In: 2018 XIV International scientific-technical conference on actual problems of electronics instrument engineering (APEIE), Novosibirsk, pp 281–284. https://doi.org/10.1109/apeie.2018.8545299
2. Annisa NN, Sensuse DI, Wilarso I (2018) Knowledge base model for call center department: a literature review. In: 2018 International conference on information technology systems and innovation (ICITSI), Bandung-Padang, Indonesia, pp 242–247. https://doi.org/10.1109/icitsi.2018.8696042
3. Bloomfield R, Lala J (2013) Safety-critical systems. The next generation. In: IEEE security & privacy, vol 11, no 4, pp 11–13
4. Mary GMJ, Munipriya P (2011) Health monitoring of IT industry people. In: 2011 3rd international conference on electronics computer technology, Kanyakumari, pp 144–148. https://doi.org/10.1109/icectech.2011.5941875
5. Rashid O, Qamar AM, Khan S, Ambreen S (2019) Intelligent decision making and planning for call center. In: 2019 International conference on computer and information sciences (ICCIS), Sakaka, Saudi Arabia, pp 1–6
6. Dai T, Li J (2015) Staff scheduling in call center considering global service quality. In: 2015 12th international conference on service systems and service management (ICSSSM), Guangzhou, pp 1–4. https://doi.org/10.1109/icsssm.2015.7170333
7. Cao Q, Qian Z, Lin Y, Zhang WJ (2015) Extending axiomatic design theory to human-machine cooperative products. In: 2015 IEEE 10th conference on industrial electronics and applications (ICIEA), Auckland, pp 329–334
8. Bentley TA, Teo STT, McLeod L, Tana F, Bosua R, Gloet M (2016) The role of organisational support in teleworker wellbeing: a socio-technical systems approach. Appl Ergon 52:207–215. https://doi.org/10.1016/j.apergo.2015.07.019
9. Zhong RY, Xu X, Klotz E et al (2017) Intelligent manufacturing in the context of industry 4.0: a review. Engineering 3(5):616–630
10. Xu MP, Wang J, Yang M, Wang W, Bai Y, Song Y (2017) Analysis of operator support method based on intelligent dynamic interlock in lead-cooled fast reactor simulator. Annals of Nuclear Energy, vol 99, pp 279–282
11. Zhirabok AN, Kalinina NA, Shumskii AE (2018) Technique of monitoring a human operator's behavior in man-machine systems. J Comput Syst Sci Int 57(3):443–452

12. Cacciabue PC (2014) Human error risk management for engineering systems: a methodology for design, safety assessment, accident investigation and training. Reliab Eng Syst Saf 83 (2):229–269
13. Dul J et al (2012) A Strategy for human factors/ergonomics: developing the discipline and profession. Ergonomics 55(4):377–395
14. Havlikovaa M, Jirglb M, Bradac Z (2015) Human reliability in man-machine systems. Procedia Eng 100:1207–1214
15. Zhang Y, Sun J, Jiang T, Yang Z (2019) Cognitive ergonomic evaluation metrics and methodology for interactive information system. In: Ahram T (ed) Advances in artificial intelligence, software and systems engineering, AHFE 2019, AISC 965, pp 559–570. https://doi.org/10.1007/978-3-030-20454-9_55
16. Burov OY (2017 May 15–18) ICT for performance assessment of emergent technologies operators. In: Proceedings of the 13th international conference on ICT in education, research and industrial applications. Integration, harmonization and knowledge transfer Kyiv, Ukraine, vol 1844. CEUR-WS, pp 127–138
17. Lavrov E, Pasko N, Siryk O (2020) Information technology for assessing the operators working environment as an element of the ensuring automated systems ergonomics and reliability. In: 2020 IEEE 15th international conference on advanced trends in radioelectronics, telecommunications and computer engineering (TCSET), Lviv-Slavske, Ukraine, pp 570–575. https://doi.org/10.1109/tcset49122.2020.235497
18. Adamenko AN, Asherov AT, Berdnikov IL et al (1993) Information controlling human-machine systems: research, design, testing. Reference book Gubinsky AI, Evgrafov VG (eds). Mashinostroenie, Moscow, Russia. (In Russian)
19. Grif MG, Sundui O, Tsoy EB (2014) Methods of desingning and modeling of man–machine systems. In: Proceedings of international summer workshop computer science, pp 38–40
20. Lavrov E, Pasko N (2018) Automation of assessing the reliability of operator's activities in contact centers that provide access to information resources. In: Proceedings of the 14th international Conference on ict in education, research and industrial applications. Integration, harmonization and knowledge transfer, Kyiv, Ukraine, vol 1, pp 445–448
21. Lavrov EA,Paderno PI, Burkov EA, Siryk OE, Pasko NB (2020) Information technology for modeling human-machine control systems and approach to integration of mathematical models for its improvement. In: 2020 XXIII international conference on soft computing and measurements (SCM), St. Petersburg, Russia, pp 117–120. https://doi.org/10.1109/scm50615.2020.9198791
22. Lavrov E, Pasko N, Siryk O, Kisel N, Sedova N (2020) The method of teaching IT students computer analysis of ergonomic reserves of the effectiveness of automated control systems. In: E3S Web of Conferences, vol 166. https://doi.org/10.1051/e3sconf/202016610017

Prometheus (PN-2003016) Digital Medical Device Collaborative E-Training and Cognitive Ergonomics to Integrate AI and Robotics in Organ Transplantations

Constantinos S. Mammas and Adamantia S. Mamma

Abstract Computer Assisted Collaborative E-Training integrating Cognitive Ergonomics of AI and Robotics to interactively train specialists in the procurement phase of Organ Transplantation (OT) is dealt. It is analyzed the clinical simulating collaborative training session among 10 specialists in OT that took place in the Megaron Hall of Athens trial on 26.02.2019 training them interactively how cognitive ergonomics integrated with AI promote the remote evaluation of the grafts and the pre- and post-grafting and pre-transplant decision making and planning so as to minimize the rates of the damaged organs focusing on infections, by using the Prometheus digital medical device (pn:2003016) integrated with the Robotic framework "Stamoulis" Rb. The results showed that AI and Robotic based cognitive ergonomics integration in OT interactive e-training enhances significantly operational maneuverability and inter-operability to remotely and with the minimum errors: a. Make accurate diagnosis and minimize damaged or diseased, with infection, trauma and cancer, grafts, b. Intensify an international clinical surveillance ergonomic system development in the procurement phase of OT, and c. Integrate ergonomically robotics in OT with the minimum errors in the clinical process.

Keywords Organ transplantation · Remote diagnosis and decision making · Collaborative e-Learning and e-Training cognitive ergonomics · AI · Big data analytics · Robotics

C. S. Mammas (✉) · A. S. Mamma
State Scholarship Foundation, Fellowships of Excellence for Postgraduate Studies in Greece-Siemens Program 2014–16, Athens, Hellas, Greece
e-mail: csmammas@med.uoa.gr

D. Sumpor et al. (eds.), *Proceedings of the 8th International Ergonomics Conference*, Advances in Intelligent Systems and Computing 1313,
https://doi.org/10.1007/978-3-030-66937-9_15

1 Introduction

Our project analyzes the e-training technology, process, ergonomics and standards of the remote, specialized, multidisciplinary and personalized evaluation of the Grafts (G) in Organ Transplantations (OT) integrating Prometheus digital medical device (pn 2003016) with AI, Big Data analytics and a Robotic framework. In that context, was given an emphasis on the remote digital and microscopic microbiologic diagnosis and evaluation and accordingly on their pre-, post-grafting and pro-transplant decision making and treatment (Prometheus pn 2003016 digital medical device technology, method, process, service and standards integrated with AI empowered Big Data analytics technology and method, and a Robotic framework called Stamoulis Rb as applied in the Megaron Hall of Athens trial on 26.02.2019 [1–5]. Currently, within the experience of the international lockdowns in 2020 the whole project seems to be the right solution in health care and specifically in OT by providing not only: (a) experts' approach for continuous training in OT for and from where ever, (b) acceleration of the diagnostic process and decision making, (c) evidenced based decisions, and (d) significantly lowering of the costs in OT, but mainly building hands-on experience and skill in (e) the remote microbiologic examination of the donors and the recipients and of the grafts in OT for accurate and remote diagnosis and pre-grafting and pre-transplant decision making and planning.

In the context of that collaborative e-training oriented session- using Prometheus (pn 2003016) digital medical device technology and method and its innovative training networks for collaborative e-learning and e-training—the invited specialists and the registered members used the official digital platform the program of Excellence 2014–16-Siemens program (www.prbite.wordpress.com), interactively and remotely, and trained on using the abovementioned digital medical device, technology and method aspects and functions (Telecommunications, Tele-radiology, Tele-pathology, Tele-cytology, Tele-microbiology, Tele-molecular biology). Also, they implemented them in the simulating clinical training practices and in the geographic and in the proper timing strategies for pre- and post-grafting remote evaluation of the Grafts and decision support and making, in OT so as to: (1) reduce the damaged and diseased organs, (2) enhance safety and improve outcomes in Liver (LT), Pancreas (PT), Renal (RT), Heart (HT), Lung (LnT) and Uterus (UT) transplants by accurate pre-transplant operative planning integrated with AI and Big Data analytics and computing [1, 3, 4, 6, 7].

The Prometheus (pn-2003016) digital medical device based technology, method, process, service, clinical and training standards analysis and ergonomics assisted collaborative e-training and e-learning applied on a simulating procurement phase based transplant scenario in an officially scheduled trial in the Megaron Hall of Athens trial on 26.02.2019 (program of Excellence 2014–16-Siemens program for Greece). Regarding the process participants applied tele-communications (TC) in the simulating Coordination Process (CP), Tele-radiology (TRE), Tele-pathology (TPE), Tele-cytology (TCE), Tele-microbiology (TME), Tele-genetic biology

(TGBE) for the remote evaluation of the abdominal (liver, kidneys, pancreas, uterus) and thoracic (heart and lung) Grafts between the donor hospital (DH) and the Recipient Hospital (RH). In addition, the interactive collaborative e-training session in the clinically simulating procurement phase of OT on 26.02.2019 integrated with AI empowered Big Data analytics and computing and a Robotic framework including: (1) Development of two training prototypes of Prometheus (pn 2003016), the two Exp.-TS (Table 1) [6, 8–12], (2) TC infrastructure among DH, National Transplant Organization (NTO) and RH based coordinators and completion and transference of the medical records of the deceased donor on the electronic space of Prometheus (pn 2003016), (3) Prometheus (pn 2003016) digital medical device based Retrospective and Real Time Projection (RTP) and TC, TRE and TPE (including TCE and TME) of the Grafts in the DH from specialists based in the RH or else, assessing the diagnostic accuracy of injury-trauma, inflammation-infection or cancer of the Grafts [4, 13–16]. According to the didactic transplant scenario on 26.02.2019, the potential donor was a young woman of 20 years who was the victim of a severe accident three days ago. Brain death had been confirmed by a neurologist and her family consented for multi-organ donation. Regarding the experimental process seven prospective and retrospective participants (n = 7 specialists: n1 = Surgeon of the Grafting Team, n2 = Transplant Surgeon, n3 = Radiologist, n4 = Pathologist, n5 = Cytologist, n6 = Microbiologist, n7 = Cardiologist) allocated in different points which connected remotely in the internet and participated in a Prometheus (pn 2003016) digital medical device based and AI and Big Data analytics clinical simulating and interactive collaborative e-training process [17–19].

In fact, Prometheus (pn-2003016) digital medical device based technology, method, process, architecture and service integrated with AI and Big Data analytics

Table 1 Comparison of the modules between OTE-TS and Prometheus Exp.-TS (pn 2003016) integrated with AI, Big Data analytics and Computing and Robotic framework (pn 34931)

Modules	OTE-TS	Prometheus Exp.-TS
Medical record process	+	+
Examinations results	+	+
Capture/imaging	+	+
DICOM and PACS	+	+
Real-time tele-conference	+	+
Chat and whiteboard	+	+
Tele-secretary facilities	+	+
Tele-mentoring facilities	+	+
Telecommunication net	ISDN based	Internet based
Virtual slide integration	–	+
Cloud computing	–	+
AI	–	+
Big data analytics	–	+
Robotic-automation integration	–	+

and Robotic framework in the procurement phase of OT, can be seen as a triangle (including the donor, the coordinator and the transplant surgeon) as the didactic base for the remote evaluation of the donor and of the abdominal and the thoracic Grafts for pre-grafting and pre-transplant instant remote macroscopic and microscopic evaluation and decision making. Also, for preoperative planning, so as to empower trainees, students and other responsible professionals in transplant, to make remotely valuable decisions by interactive e-learning and e-training of the up to date clinical guidelines and contemporary technology in OT. So, participants had the chance to learn and practice the right processes and develop skills for valuable clinical services in OT to confront clinical transplant complexity at distance and in low cost and improve significantly therapeutic outcomes on a national and international level. Moreover, they developed the necessary clinical thoughts and hands on practice by using high technology to build-up the special skills needed to diagnose remotely the macroscopic and microscopic lesions of the damaged grafts and make decisions about their use or reduction in OT [1, 2, 4, 20–24].

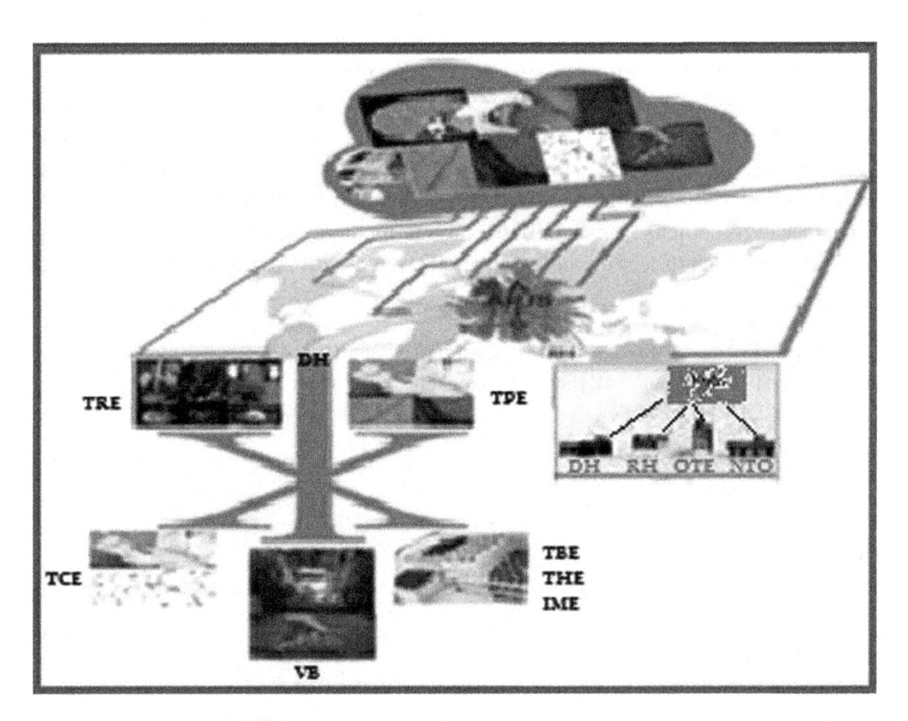

Fig. 1 The remote evaluation of the Grafts as applied on 26.02.2019 after virtual benching (VB) upon OTE NETWORKS, among transplant coordinators (NTO), grafting and transplant surgeons and other specialists between the simulating Donor Hospital (DH) (Aretaieion University Hospital of Athens) from where structured Data would be processed in the integration of Prometheus Digital Medical Device (pn 2003016) with the Cloud and AI empowered process consisting the Robotic framework Stamoulis Rb (pn 34931) and sent to the Recipient simulating Hospital (RH) (Hippocrateion University Hospital of Thessaloniki)

The interactive collaborative e-training process on 26.02.2019 started by a retrospective projection of the results and data that had been taken and archived from two departments of Pathology and of Radiology respectively of the medical school of Athens which simulated the DH. The results and the used digital data had been archived in the innovative networks of Prometheus (pn 2003016) digital medical device technology and method in the collaborative e-training in the Megaron Hall of Athens trial on 26.02.2019 (Cloud computing AI and Big Data analytics and Robotic framework integration). According, to the scenario the "Paros" conference room of the Hilton-Athens simulated the RH for the CP and for the remote macro- and micro-collaborative evaluation (TRE, TPE, TCE and TME). Participants allocated to 7 points for remote macro- and/or micro-diagnosis i.e., Point 1 = Department of Pathology in the DH, Point 2 = RH, Department of Immunology for CP, Point 3 = Department of Pathology for TPE, Point 4 = Department of Radiology for TRE, Point 5 = Department of Microbiology for TME, Point 6 = A Department of Cardiology in the RH, Point 7 = A Department of Cytology in the RH for TCE (Fig. 1).

2 Paper Content and Technical Requirements

Given that the ergonomic impact of the electronic space (ES) for the remote macro- and micro-examinations diagnosis of the Grafts, had been already evaluated on the digital medical device Prometheus (pn 2003016) in the context of previous experimental simulations, in that collaborative e-training session on 25.11.2018, that material used retrospectively. In detail, it referred to the ergonomics of Digital Macroscopy and Microscopy (DM) in Tele-medicine and TPE, in TCE and in TME, for the remote macroscopic and microscopic diagnosis of injury/trauma, inflammatory, infectious and/or neoplastic lesions of the Grafts for decision making about their use and treatment. The simulating experimentation for the collaborative e-training on 26.02.2019 also included: (1) The development of an experimental digital medical device similar to Prometheus pn 2003016 and to OTE-TS, the Exp.-TS [5, 9, 10], (2) Integration of the Exp.-TS with the Virtual Slide system (VS) for digitalization of images for TPE or TCE applying DM, AI analytical method and Robotics in the process (Table 1), (3) Simulation of static or dynamic TPE and TCE for microscopic diagnosis of injury/trauma, inflammatory, infectious and/or neoplastic lesions (Fig. 1) [11–13].

2.1 Designing for Human Factors: The Technology for Remote Evaluation of the Grafts in OT

According to the experimental protocol on 26.02.2019, the Prometheus (pn 2003016) digital medical device integrated with AI and Big Data analytics and computing based simulating clinical session, while the data sourced from a Robotic framework integrated with Cloud archives, from which Prometheus (pn 2003016) accumulated them from a concrete path for: (1) retrospective and/or real time remote diagnosis of the possible Grafts' lesions, applying telemedicine and remote visualization standards and (2) computations of the requested clinical and financial indicators (number of transplantations, number of defected Grafts, cost of post-ponements in the total Transplant process because of the damaged Grafts, the total cost of the damaged or defected Grafts on a European level according to the data archived in the Robot) i.e., (1) clinically—for the remote evaluation of the grafts and decision making—and (2) financially for financial indicators (total cost and amount losses because of the damaged organs OT postponements) instant analysis and computations and decision making about the quality of the transplant projects in each country on a European level. The results depicted, instantly, on tables and graphs for each year, and for each Big Data cost analytics for each graft type (Heart, Lung, Liver, Pancreas, Kidneys). The results depicted instantly and used by the participants in a table and in a graph for Big Data analytics and computing process from the participant users on 26.02.2019 (Fig. 2).

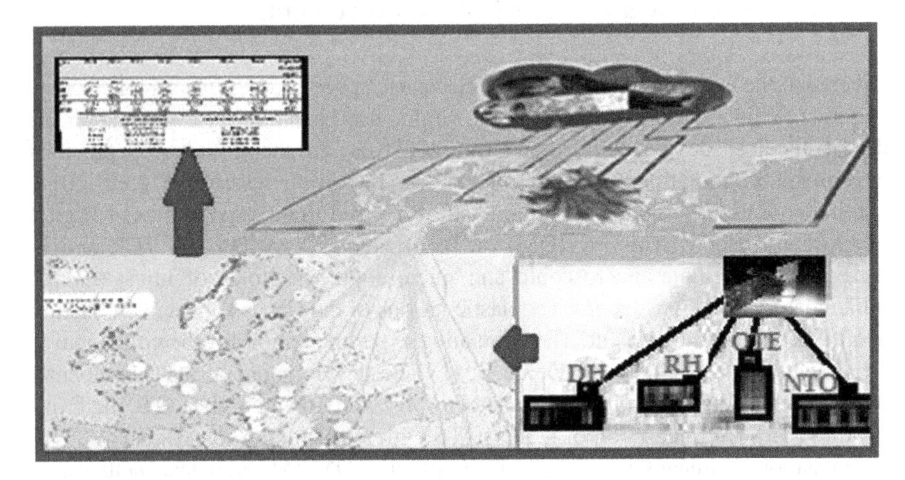

Fig. 2 The Prometheus (pn 2003016) digital medical device, AI and Big Data integrated collaborative e-training in the Megaron Hall of Athens trial on 26.02.2019, connected with partners allocated to all European transplant centers—from which the digital images and the numbers of transplantation cases and that of the damaged organs had been received by each NTO —and after formation of structured Data, would be sent to the Cloud from the Robotic framework Stamoulis Rb in the DH for AI empowered retrospective and real time Big Data analytics and computing in the RH and the NTO (pn 34931)

2.2 Prometheus (pn 2003016) Based Collaborative E-Training for the Remote Evaluation of the Grafts in OT Integrating AI, Big Data Analytics and Robotics

The integrated TC, CP, TRE, TPE, TCE, TME and TGBE of solid abdominal and thoracic organs i.e., PG, LG, RG, HG, LnG and UG for the minimization of the damaged organs in OT and for pre-grafting and/or pre-transplant decision making and planning had shown high sensitivity, specificity and accuracy (ranging from 90 to 98%) in the remote diagnosis of traumatic, infectious, inflammatory, neoplastic lesions. Taking into consideration that a modern Hyper-computer is equipped with accelerating modules and with new technologies and that is the perfect choice for integrated Big Data analytics and computing and CE [18, 20, 21], the corresponding transplantations data in Europe from 2011 to 2016 had been archived in the working Cloud of the Robot of the program of Excellence 2014–16-Siemens Program for Greece, as a structured material for real time Big Data analytics and computing, so that Prometheus (pn 2003016) digital medical device users could accumulate them for: (1) remote diagnosis collaborative interactive e-training, and for (2) instant Big Data computations training on 26.02.2019 [22–26].

Prometheus (pn 2003016) based Collaborative E-Training for Real Time Big-Data analytics and computing, of the damaged organs rates and their cost.

In the collaborative e-training process the material that used by the participants on 26.02.2019 based on the GODT publications data from 2011 to 2016 and archived in the Cloud. The real time Big Data financial analysis with the technology and method of the digital medical device Prometheus (pn 2003016) completed with the real time computation of the damaged organs rate in analogy to the estimated indexes in Britain's OT project and the related published data in 2010 [1, 4]. Regarding damaged organs cost analysis, Big Data analytics and instant computing took into consideration that Prometheus (pn 2003016) digital medical device received instantly the total number of transplantations from NTO or ESOT. The experimental process based on the GODT publications from 2011 to 2016. The real time Big Data financial analysis with the technology and method of the digital medical device Prometheus (pn 2003016) completed with the real time computation of the optimum damaged organs rate according to the estimation indexes in Britain in 2010 and their costs taking into consideration that the used digital medical device received on real time the total number of transplantation that would take place in a form of structured data from the National Transplant Organizations in Europe and/or from the ESOT using the Cloud technology (Fig. 2).

Experimental results analysis in a real time mode using the digital medical device Prometheus (pn 2003016) showed high feasibility for the direct computational analysis of Big-data referring to the cost of transplantation per organ, to the estimation and comparison of the referred damaged organs and to their cost (Tables 2, 3). On the other hand the accuracy of the computational analysis was really high given that a software for mathematical analysis had been incorporated in the system (Fig. 1 and Tables 2, 3). The process for the final computations of the

Table 2 The visual quality classification scale (−3, −2, −1, 0, 1, 2, 3) used by the participants and the results given by the collaborated microbiologist on 26.02.2019

Microscopic visual quality parameters	Classification
Quality	1
Sharpness	1
Brightness	1
Size of window	1
Form	1
Color	1
Final visual organization	1

Table 3 Actual organ donations performed in the European Union from 2011 to 2016

Organ	2011	2012	2013	2014	2015	2016	Total
Heart	1980	1960	2037	2146	2235	2254	12,612
Lung	1677	1756	1825	1822	1818	1916	10,814
Liver	7006	6845	7173	7390	7694	7762	43,870
Kidney	18,712	18,854	19,227	19,670	20,102	20,638	117,203
Pancreas	859	825	865	818	821	780	4968

costs of interest in this particular experimentation (Total cost per OT and losses because of the damaged organs)—using the digital medical Prometheus (pn 2003016)—lasted for an additional half an hour (Fig. 2 and Tables 2, 3). The abovementioned computational analysis results can be used by the Ministry of Health services or the National Transplant Organizations to intervene locally or nationally on time so as to correct several problems referring to the rate of damaged organs in OT, to organ donation and to the cost of OT in the country or to correct mistakes or destructive pricing policies of the biomedical products on time [9–11, 18]. The results of that last collaborative e-training session also showed high usability and reliability as a training device and method for learning the technology, the method and the clinical criteria, the process and the ergonomics of the remote evaluation of the Grafts and for instant computations of the clinical risks sourced from the damaged organs and their correction in OT—including the financial ones —shaping the AI enhanced Big Data analytics and computing collaborative e-learning and e-training in OT integrated ergonomically with a robotic framework (Stamoulis Rb).

Prometheus (pn 2003016) based collaborative e-training: The post-grafting TME of the donor's grafts (prospectively on 26.02.2019).

The simulating process of the remote evaluation of the right renal Graft for the diagnosed cyst liquid included pre-grafting RTP TME of the collected liquid for direct gram stain. The remote multidisciplinary, specialized and personalized microbiological diagnosis performed prospectively by a specialist Microbiologist located in the RH at the time of experimentation on 26.02.2019 [9, 11, 22, 23, 25].

The diagnosis included RTP TME especially with remote microbiologic microscopic analysis of the collected samples and the culture of the collected liquid of the cyst of the grafted right kidney and of the collected donor's urine as well. The microbiologist's report is as follows: (1) With regard to the remote microscopic visual quality of the RTP TME—using Prometheus digital medical device (pn 2003016) on 26.02.2019—for both the urines and the cyst's liquid gram stain and cultivation plate, microscopic examination, the sample was acceptable for remote diagnosis and the classification score was 1 for all visual parameters (Table 2), (2) The Donor's urine—according to the simulating clinical scenario—that collected directly from the ureter of the graft and stained with a gram stain showed—after RTP TME i.e., remote microscopic analysis is possibly infected with gram(-) bacteria, indicating also 12–15 neutrophils (pof). Additional examination of the examined cystic liquid of the donor's right kidney Graft showed—after RTP remote microscopic analysis—a possible infection with gram(-) bacteria and with 10–12 neutrophils (pof), (3) Cultivation of the donor's urine and the right's kidney cyst's liquid—and the remote microscopic analysis of the plates after RTP TME i.e. using Prometheus (pn 2003016)—by the part of the participant microbiologist showed: (1) Proteus aureus species in the donor's urine and (2) *Escherichia coli* species in the grafted right kidney's cyst liquid. Initially, the microbiologist found the information significant to change decisions about the continuation of the transplant process and about the prophylactic or therapeutic anti-infection scheme for the recipient in the RH in case the renal Graft would be decided to be sent to the RH, (4) The results recorded and then a cooperator consultant medical physician and a specialist in transplant infections (Consultant Oana Ramayana from Paris)—training in the technology and method of Prometheus (pn 2003016) as a participant—remotely requested to advice upon the abovementioned results for indication or contra-indication of sending the Graft from the DH to the RH for the virtual renal transplant on 26.02.2019.

Prometheus (pn 2003016) based collaborative e-training on 26.02.2019: Real time Big-Data analytics and computing of the damaged organs in OT.

In the experimental process on 26.02.2019, the material used among participants based on the GODT publications data from 2011 to 2016 working Cloud. The real time Big Data financial analysis with the digital medical device Prometheus (pn 2003016) completed with the real time computation of the optimum damaged organs rate according to the estimation indexes in Britain in 2010 and their costs taking into consideration that the used digital medical device received on real time the total number of transplantations that would take place in Europe, in a form of structured data from each National Transplant Organizations in Europe and/or from the ESOT's Cloud technology on a weekly base (Figs. 1, 2). The accuracy of the abovementioned computational analysis was really high, given that a software for mathematical analysis had been incorporated in Prometheus (pn 2003016) so as to calculate instantly an project the Big Data computations in the Cloud (Figs. 1, 2 and Tables 2, 3, 4, 5). The total time for preparation of the digital data and their projection to the Cloud of the structured Data for final computations for both types of costs was less than half an hour, on 26.02.2019. Correspondingly, the process for the

Table 4 Computations of the expected damaged organs as computed by big data analytics in the collaborative e-training on 26.02.2019

Organ	Expected damaged organs
Heart	1261
Lung	1081
Liver	3510
Kidney	12,892
Pancreas	696

Table 5 Total expenditures in OT- and loses in case of damaged organs in Europe from 2011 to 16 as computed by big data analytics in the collaborative e-training by the participants on 26.02.2019

Total cost of the performed OT (2011–16) in US dollars	Estimated cost of the damaged organs in the performed OT (2011–16) at a minimum in US dollars
$258299e10	$1.609291e9
$6.068817e9	$901887600
$531738e10	$4.229068e9
$391855e10	$9.868493e9
$1.437739e9	$407376000

final computations of the costs of interest in this particular experimentation (Total cost per OT and losses because of the damaged organs)—using the digital medical Prometheus (pn 2003016)—lasted for an additional half an hour (Figs. 1, 2 and Tables 2, 3, 4, 5). The abovementioned computational analysis results can be used by the Ministry of Health services or the National Transplant Organizations to intervene locally or nationally on time, so as to correct several problems referring to the rate of damaged organs in OT, to organ donation rate and to the cost of OT or to correct mistakes and/or destructive pricing policies of the biomedical products, on time [23, 25]. On the other hand they may be used as accurate and real time tools on a weekly base to make decisions about the biomedical technology products in OT expenses and pricing for the benefit of the population and the state [1, 3, 11]. The main drawback of this experimental study is that integration of Big-Data computational analysis on 26.02.2019 based on a six years framework of clinical organ transplantations on a European level simultaneously and not on a weekly level and that computations based on the 2011 prices, in US Dollars. The feasibility and reliability comparative analysis referred mainly to the evaluation of the use of Prometheus (pn 2003016) digital medical device technology and method, integrated with AI and Big Data analytics as a training device and the above mentioned limitations didn't have any impact in our simulating collaborative e-training process. Another drawback was that there was non-parallel comparative computational process using the Hyper-computer ARIS-IBM. However, the reliability and accuracy of a modern Hyper-computer considered by definition as the ideal one. In addition, we are developing new projects for testing it, integrating robotic automation process [12, 14].

3 Conclusion

This new training dimension of Prometheus (pn 2003016) digital medical device with integration of AI and Big Data analytics and computing and a Robotic framework for e-collaborative training in the procurement phase of OT on 26.02.2019 coincides with the specialist transplant surgeon's attestation—after experimentation on 28.06.2016—about its possible clinical contribution in the organs' safety in OT [3, 21]. In the same context, participants had underlined the significance of the accumulated clinical experience and the continuous and intensive training from the part of the transplantation centers to benefit from our innovative integrated system and ergonomics in OT [22, 23, 25]. On the other hand, modern technological integrations such as the intersection of knowledge, CE methods AI and Big Data analytics and Robotics create an important new framework for driving new in-sights remotely, continually and on real a time mode reference to every day transplantations for making evidenced decisions and for computing the number of damaged organs in OT performed on a national, European and Global level weekly taking into consideration the current conditions of pandemics and lockdowns world-wide. The same counts for the real time computation of the concurrent national or super-national expenditures. As analyzed the amounts spent or wasted because of the damaged organs in OT are tremendously high and the lack of evidenced decisions may cause severe financial losses on a national level. Hence, integration with Hyper-computers makes possible the ergonomic integration an elaboration of AI empowered Big-Data analytics and computing in OT to: a. Minimize damaged or diseased grafts from donors infected with viral diseases, b. Teach semantic technologies, c. Reduce time, cost and maximize outcome and enhance personalization in OT [18, 25, 26]. To the same direction, the integration of the Robotic automation process enriched the collaborative e-training and e-learning material and process for OT on 26.02.2019 and the results showed that may be feasible and reliable even for other medical or surgical fields fulfilling automatically and reliably, all the above mentioned complex clinical, technological and training objectives and standards to optimize quality by deeper ergonomics in OT [27].

References

1. Hippocrates (1972) Ancient medicine (Trans. Jones WHS; Loeb Classical Library). Harvard University Press, Cambridge
2. Souba WW (1997) The commercialization of American medicine: are we headed for curing without caring? J Surg Res 67:1–3
3. Mammas CS (2009) On the poor patient. Periopton.gr, Athens. (In Hellenics)
4. Paraskevaidis C (1998) On the moral perspectives of the financially-centered health. Autopublication, Athens. (In Hellenics)
5. Pellegrino ED, Thomas DC (1996) The christian virtues in medical practice. Georgetown University Press, Washington DC

6. Trichopoulos D, Trichopoulou A (1986) Diseases prevention. EKPA, Athens. (In Hellenics)
7. Priester R (1991) Distributing limited health care resources. Center for Biomedical Ethics, University of Minnesota, Minessota
8. Sandel JS (2012) What money can't buy-the moral limits of markets. Allen Lane, London
9. TO BHMA (2012) 90 Years. In the era of the collapse of illusions. Labrakis PO, Athens. (In Hellenics)
10. Kousoulis L (2016) Day after day Hellas is sinking in a night without dawn. Pronews 2:62. (In Hellenics)
11. Friend P. Organization of organ retrieval-bitter lessons and ambitious plans. http://vincentbourquin.files.wordpress.com/2010/11/friend-Hammersmith-11-10-10.pdf. Last accessed 2012/07/17
12. Raghupathi W, Raghupathi V (2014) Big data analytics in healthcare: promise and potential. Health Inf Sci Syst 2(3):1
13. Wilson KM, Helton WS, Wiggins MW (2013) Cognitive engineering. Wiley Interdiscip Rev Cogn Sci 4(1):17
14. Aoun MA (2016) Advances in three hyper-computation models. EJTP 13(36):169
15. Gualtieri M. True streaming analytics platforms for real-time everything. Available from https://go.forrester.com/blogs/16-04-16-. Last accessed 2018/02/03
16. Panahiazar M, Taslimitehrani V, Jadhav A, Pathak J (2014) Empowering personalized medicine with big data and semantic web technology: promises, challenges, and use cases. In: Proceedings of the IEEE international conference on big data, IEEE, pp 790
17. Mammas CS, Mandellos G, Economou GP, Lymberopoulos D (2001) Structuring expert-leaded medical protocols for telemedicine systems. In: Proceedings of the 23rd annual international conference of the IEEE engineering in medicine and biology society proceedings, IEEE, Constantinople, pp 3529
18. Karavatselou E, Economou GP, Chassomeris C, Daneli V, Lymperopoulos D (2011) OTE-TS-a new value-added telematics service for telemedicine applications. IEEE Trans Inf Technol Biomed 5(3):210
19. Morgan RH (2001) Computer network security for the radiology enterprise. Radiology 220(2):303
20. Mammas CS. The remote evaluation of pap-test applying tele-cytology in the context of uterus graft remote evaluation and in the remote prevention of cervical cancer. http://www.lib.uoa.gr/fileadmin/user_upload/I.K.Y_Papanikolaou.pdf. Last accessed 2015/04/11
21. Matesanz R (2011) NewsLetter Transplant 17
22. Matesanz R (2012) NewsLetter Transplant 18(1)
23. Matesanz R (2013) NewsLetter Transplant 19(1)
24. Matesanz R (2014) NewsLetter Transplant 20(1)
25. Matesanz R (2015) NewsLetter Transplant 21(1)
26. Matesanz R (2016) NewsLetter Transplant 22(1)
27. Mammas CS, Mamma AS (2017) Innovative training networks for interactive e-learning and application of ergonomics of the remote evaluation of the grafts in organ transplantation. In: Proceedings of the 49th annual conference of nordic ergonomics and human factors society on proceedings. Lund University, Lund, pp 114

Prometheus I (PN 1008239) Digital Medical Device Integrated with AI and Robotics Cognitive Ergonomics in Breast Cancer Prevention

Constantinos S. Mammas and Adamantia S. Mamma

Abstract The necessity, technology, process, service, and ergonomics of the clinical remote massive public education, primary, secondary and tertiary breast cancer prevention upon innovative networks for specialists and citizens in the context of the clinical prevention operations of the tele-medicine and cloud technology based novel mobile clinical unit (MCU) called "Macedonia" has been studied, tried and standardized in the program of Excellence 2014–16. It is analyzed the clinical prevention operability, maneuverability and reliability of the above mentioned technology and method empowered with cognitive ergonomics of AI and a robotic framework the "Charicleia" Rb integrated with Prometheus I (pn 1008239) digital medical device all integrated with (MCU) in the tenth and the eleventh clinical diseases prevention action in Promyrion village of the Municipality of South Pelion Mountain and in the Parish of Agioi Anargyroi of Volos on 05-06.06 and in Demene village of Municipality of Aisonia on 07.07.2020 for remote clinical remote massive public education, primary, secondary and tertiary breast cancer prevention. Ergonomic Integration of (MCU) operations enriched with AI and Big Data analytics and Robotics may optimize: (1) breast cancer prevention status, (2) quality of healthcare especially in developing countries.

Keywords Breast cancer prevention · Remote diagnosis and decision making · Cognitive ergonomics · AI · Robotics

1 Introduction

Human factors (HF) engineering is a discipline concerned with the design of tools, machines, and systems that take into account human capabilities, limitations, and characteristics [1, 2]. The impact of HF in terms of reliability analysis of the clinical

C. S. Mammas (✉) · A. S. Mamma
State Scholarship Foundation, Fellowships of Excellence for Postgraduate Studies in Greece-Siemens Program 2014–16, Athens, Hellas, Greece
e-mail: csmammas@med.uoa.gr

D. Sumpor et al. (eds.), *Proceedings of the 8th International Ergonomics Conference*, Advances in Intelligent Systems and Computing 1313,
https://doi.org/10.1007/978-3-030-66937-9_16

operations of the new Mobile Clinical Unit "Macedonia" (MCU) integrated with Prometheus I (pn 1008239) digital medical device, AI and Big Data analytics and computing and a Robotic framework for remote primary, secondary and tertiary general for cancer and specific for diabetes related complications primary, secondary and tertiary prevention, is the topic of our high technology operational and high demanding quality massive diseases prevention service for Greece and for developing countries (DC). In this context, the compact MCU "Macedonia", cloud computing technology and method for the remote, specialized, multidisciplinary, holistic and personalized medical process and service, integrated with AI, Big Data analytics and a Robotic framework is assessed. On a clinical level the above mentioned used for remote mass health education and for general for cervical and breast cancer for women and for prostate cancer for male population and for large bowel cancer for both sexes and special for diabetes related complications with an emphasis on diabetic foot primary, secondary and tertiary prevention of amputations in Greece in the village of Promyrion of the Municipality of the South Pelion and in the Parish of Agioi Anargyroi in the City of Volos on 05.06.2020 and on 06.06.2020 respectively. Then into Demene village of the Municipality of Aisonia on 07.07.2020 [3]. Referring to cancer prevention it is well known that Breast Cancer (BC) is the first cancer in the female population and Cervical Cancer (CC) is the second most common cancer world-wide and the leading cause of cancer related deaths among women in developing countries (DC) [4–6]. In this context, the novel Hellenic clinical mobile transportation unit (MTU) or mobile clinical unit (MCU) called "Macedonia"—Two prototypes constructed in the Program of Excellence 2014–16-Siemens program for Greece—has been augmented and standardized with the operational ergonomics of Prometheus I pn 1008239 digital medical device technology and method and it has been evaluated from clinically, technologically, since 2013 [7, 8]. The above mentioned counted not only for Greece but for DCs as well as during the international lockdowns in 2020 [7].

2 Paper Content and Technical Requirements

The MCU Macedonia supported by Prometheus I (1008239) and integrated with AI and Big Data analytics and computing based Specialized, Multidisciplinary, Holistic and Personalized Prevention Operation for collaborative e-training of the participants for general diseases, for diabetes related complications and for CC, BC, PC and LBC prevention as performed in the village of Promyrion of the Municipality of the South Pelion and in the neighborhood of Agioi Anargyroi Parish in the City of Volos on 05.06.2020 and on 06.06.2020 respectively. Then, into Demene village of the Municipality of Aisonia on 07.07.2020. By experimental simulation, the clinical ergonomic impact of the electronic space (ES) for remote macro- and micro-examinations has been already evaluated on the digital medical device Prometheus I pn 1008239 first experimentally in terms of the ergonomics of Prometheus I pn 1008239 Digital Macroscopy and Microscopy (DM) in

Tele-medicine and Tele-pathology (TPE), in Tele-cytology (TCE) and Tele-microbiology (TME), for macroscopic and microscopic inflammatory, infectious and/or neoplastic lesions feasibility, reliability and accuracy in 2013 and presented in 2013–14. In continuation, simulating experimentation for the use of Prometheus I pn 1008239 digital medical device for collaborative e-training among participants and experts on 26.07.2018 included: (1) The Development of an OTE-TS similar digital device (PROMETHEUS I pn 1008239) [5, 9, 10], (2) Integration of the Exp.-TS with microscopic infrastructure for digitalization for TPE or TCE applying DM (Table 1), (3) Integration of the Exp.-TS with the Robotic framework "Charicleia" Rb armed it for further examinational automation and remote visualization (Table 1) (4) Simulation of static or dynamic TPE and TCE for remote macroscopic, microscopic and radiological diagnosis of inflammatory or neoplastic lesions especially of the Breast (Fig. 1) [11–13]. According to the experimental protocol in that MCU Macedonia and Prometheus pn 1008239 digital medical device based clinical in the village of Promyrion of the Municipality of the South Pelion and in the neighborhood of Agioi Anargyroi in the City of Volos on 05.06.2020 and on 06.06.2020 (totally 13,000 citizens) respectively. Then into Demene village of the Municipality of Aisonia (2000 citizens) on 07.07.2020.

The data sourced from the 13 peripheries of Greece formed as structured Big Data and archived to the integrated private for the Program of Excellence 2014–16 official private Cloud, from which the digital medical device Prometheus I (pn 1008239) accumulated them for real time Cloud computing of the requested clinical and financial indicators: Prevention examinations computing according to the age and sex of the work force of Thessaly to compute the: total cost, cost paid by the

Table 1 Comparison of the modules between OTE-TS and Prometheus I Exp.-TS (pn 1008239) integrated with AI, Big Data analytics and a Robotic framework

Modules	OTE-TS	Prometheus I Exp.-TS
Medical record process	+	+
Examinations results	+	+
Capture/imaging	+	+
DICOM and PACS	+	+
Real-time tele-conference	+	+
Chat and whiteboard	+	+
Tele-secretary facilities	+	+
Tele-mentoring facilities	+	+
Telecommunication net	ISDN based	Internet based
Virtual slide integration	–	+
Cloud computing	–	+
AI	–	+
Big data analytics	–	+
Robotic-automation integration	–	+

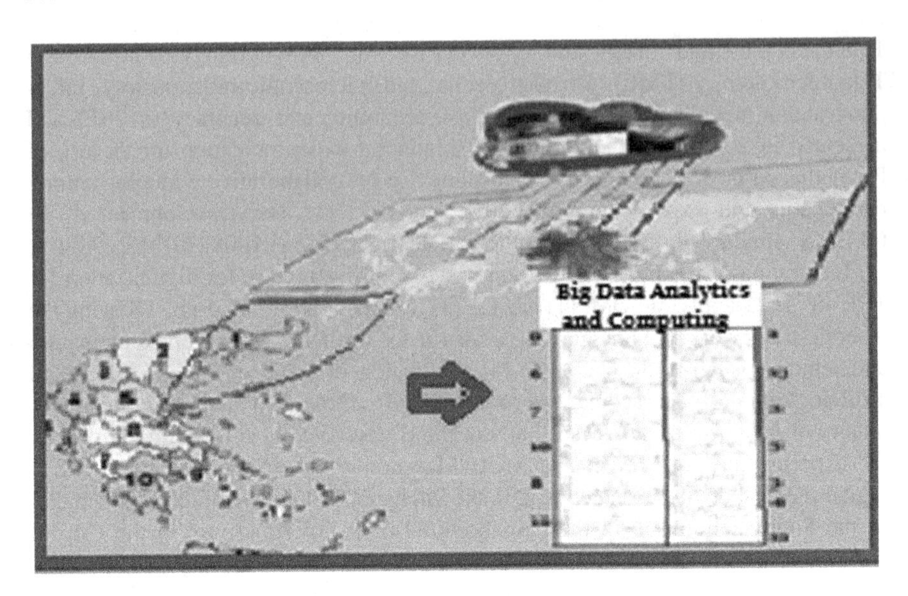

Fig. 1 The architecture and the process of the network upon which the MCU "Macedonia" integrated with Prometheus I (1008239) digital medical device, Cloud, AI, Big Data analytics and computing and the Robotic framework "Charicleia" Rb (pn 24932, pn 24933). Data would be sent from the MCU to the remote participants for AI supported computation of the annual preventive examinations of the Breast and for massive primary public health prevention according to the sex and the age as well as for all the other preventive examinations and for Big Data analytics and computing for the total work-force of the Periphery of Thessaly on 05-06.07.2020 and on 07.07.2020 (Daily press publication "Thessalia" Journal of Volos city on 18.10.2020)

citizens, cost paid by the state etc., for the periphery of Thessaly. The results depicted instantly on a table and in a graph (Fig. 1) [16, 17].

All the sixty (n = 60) participants—who were located in the three places in total —who, during the sessions, first under-took AI empowered and related to their gender and age annual prevention examination protocol for further computations shaping the remote massive public health education for cervical, breast, large bowel and prostate cancer and for special for diabetes related complications primary (diabetic foot, obesity, quit smoking, cardio-circulatory, ocular, neurologic, psychiatric), secondary and tertiary prevention. As said, then only diabetics examined clinically and in case there was a high risk patient a remote clinical examination applying the digital medical device Prometheus I (1008239) integrated with AI, Big Data analytics and the robotic framework "Charicleia" Rb. Thus the initial e-collaborative training session in primary care by AI enhanced instant computations according to the Ohio's Diseases Prevention Protocol in the USA for almost 15,000 citizens processed remotely and instantly from all the three locations that the MCU had approached [17]. Then, as above mentioned only the diabetics or patients suffering from severe—mainly chronic neuro-muscular diseases such arthritis etc. —diseases examined clinically for secondary and tertiary diseases prevention and in case there was a high risk patient a telemedicine based examination would be applied by Prometheus I pn 1008239 digital medical device in connection with the center. In these three clinical MCU sessions participated physically n = 60 citizens from which finally examined n1 = 37 patients (Females = 31 and Males = 6) while n2 = 6 were diabetics (10%). Operational Reliability of applied MCU and Prometheus I pn 1008239 integrated technology and method as well as ergonomics analyzed on three dimensions of effectiveness: (1) the medical outcome, (2) the cost-benefit analysis and (3) the fulfilment of the personalization issue. In technological terms, to achieve the above mentioned prevention operation architecture of the MCU "Macedonia" integrated with Prometheus I pn 1008239 AI and Big Data analytics and the Robotic framework Charicleia Rb infrastructure we used cloud computing (Fig. 1) [9, 14, 15, 17, 18].

2.1 *Applied Ergonomics of the MCU—Supported by Prometheus (pn 1008239) Digital Medical Device and AI and Big Data Analytics and Computing and the Robotic Framework*

Referring to the mass remote health education and primary health prevention on the 05-06.06.2020 and on 07.07.2020 clinical and technological use of the MCU "Macedonia" and Prometheus I pn 1008239 digital medical device AI empowered Big Data analytics system with a Robotic framework clinical sessions it was designed first to perform specific computations referring only to the working force of the Periphery of Thessaly—where the three places belong to. Thus, 15,000 citizens received instantly the AI empowered protocol of the annual prevention scheme for self-based computations of their annual prevention examinations according to their age and sex—we estimated, instantly, according to the Ohio's annual prevention program. Referring to the total work force of both the private and the public sector of the Periphery of Thessaly (Using official data of the 2011's condition) we estimated the total cost for performing the whole prevention examination scheme applying structured Big Data analytics and computing. Clinically, examined n3 = 31 female citizens from which n4 = 25 examined instantly also with radiological examination upon indication: (1) a non-diabetic patient who suffered from a nipple discharge and needed a remote microscopic examination for decision making and management done retrospectively for remote cytological examination by our system shaping an innovative and effective combination and integration of macroscopic and microscopic clinical, radiological and microscopic remote breast cancer prevention and decision making system, (2) another non-diabetic patient suffered from acute breast hypertrophy lesions needed further radiological and expert based informed decision making done instantly by our system also shaping an innovative and effective combination and integration of remote macroscopic clinical, radiological remote breast cancer prevention and decision making, (3) two non-diabetic n5 = 2 relatives of an examined patient with a medical history of Breast Cancer received AI empowered remote genetic counseling for Breast Cancer prevention and decision making for further examinations or the annual prevention scheme (Table 2) [17, 19].

Apart from the massive remote AI empowered remote e-public health education for the 15,000 citizens related to breast cancer prevention in few hours instantly, on 05-06.06.2020 and on 07.07.2020 as mentioned, if a remote tertiary prevention examination indicated it would be processed instantly by a AI, Big Data analytics and computing integrated robotic framework assisted process (the "Charicleia" Rb) directly to the connected tertiary health care center in a static or dynamic mode. There were four (n6 = 4) remote interventions—out of the 31 local clinical examinations for Breast Cancer Prevention. All four tertiary prevention remote examinations were accurate while the concurrent interventions leaded to a final clinical problem solving (Tables 2, 3) [9, 14, 17, 18].

None of the examined patients died (mortality 0%). All four remote diagnosis were accurate and interventions successful leading to right diagnosis or therapy or counseling of severe acute or chronic breast lesions and problems (Table 4) [17–19]. The AI empowered Big Data analytics and computing processed by Prometheus I pn 1008239 referred to the estimation of the cost of the prevention examinations for both the public and the private workers in the Periphery of Thessaly, recomputed instantly in the context of the Agioi Anargyroi of Volos city by the MCU integrated clinical operation on 06.06.2020 (Table 5).

A cost of 31.618.416.53 Euros—in 2011 prices and workforce situation in Thessaly-has to be spent for a decent diseases prevention service, annually, only in the central part of Greece, which is in reality isn't spent by the state so that the diseases prevention rate in Greece to be on a very basic level with tremendous consequences in morbidity and mortality, in the houses finances and in the national finances as well. On the other hand, the cost-effect analysis this was taken for granted on 06.06.2020 given that the comparison of the cost between: (1) the

Table 2 Breast Cancer Prevention Signs and diseases diagnosed on 05-06.06.2020 and on 07.07.2020

Sign/Disease (31/60 patients)	Parameter	N	%
Clinical examination	Yes	31	51
Radiological examination	Yes	25	41
	No	6	59
Breast lesions	Yes	2	3
Genetic counseling	Yes	2	3

Table 3 Types of e-intervention after indication for remote assessment (4/60)

Remote tertiary interventions	N	%
AI empowered e-public health education	15×10^3	
E-diagnosis and decision making	4	7
Personalized e-genetic counseling	2	3
Personalized e-cytological examination	1	2
Personalized e-prescription	31	51

Table 4 Outcome after e-examination decision making and treatment in 4 patients (4/60)

Outcome	N	%
e-Follow up and orders	31	100
Mammography after e-prescription	31	100
Reduction of Breast edema redness, warmness	1	
Reduction of Breast pain level	1	
Healed breast lesions after remote diagnosis and intervention	1	

Table 5 Cost of the preventive examinations of the workforce in the periphery of Thessally*

*	Total cost (in Euros)	Cost paid by the individuals (in Euros)	Cost paid by the State (in Euros)
Cost	31.618.416.53	4.415.018.14	27.203.398.39

*Based on the workforce and finance status of 2011

current practice in Greece and (2) the above mentioned technology process and service—based on Prometheus I pn 1008239 and the MCU—AI and Robotic Intervention had showed a significant difference favoring the second as estimated since 2015 ($p = 0.048$).

3 Conclusion

Reliability analysis of the clinical operations of the new MCU, integrated with Prometheus I pn 1008239 digital medical device, AI and Big Data analytics technology and the Robotic framework "Charicleia" Rb method and standards on 05-06.06.2020 and on 07.07.2020 showed high feasibility, clinical reliability and cost-effectiveness for the participants citizens and patients in both local and remote health care education, primary, secondary and tertiary diseases prevention [18, 19]. Among the limitations of the project it is that analysis is based on both prospective and retrospective assessment and decision making and that the remote Health Education followed the local clinical examinations program [17, 18]. Taking into consideration the above mentioned and the covid-19 pandemics and lockdowns internationally the MCU "Macedonia" integrated with Prometheus I pn 1008239 digital medical device AI, Big Data analytics and the robotic framework "Charicleia" Rb technology and method for continuous clinical processes and services is needed to cover-up the whole country and internationally other epidemiologically DCs—in terms of their healthcare condition. Ergonomically this may optimize their: (1) diseases prevention status and currently during the lockdowns because of the Covid-19 epidemiological impact, (2) quality of healthcare in total and (3) their economies, (4) clinical prevention activity during the continuous lockdowns because of epidemics or pandemics [19].

References

1. Martorell S, Soares C, Barnett J (2014) Safety, reliability and risk analysis: theory, methods and applications. CRC, United States
2. Krupinski E (2014) Human factors and human-computer considerations in tele-radiology and tele-pathology. Healthcare 2:94
3. Tentolouris N et al (2012) Type 2 diabetes mellitus is associated with obesity, smoking and low socioeconomic status in large and representative samples of rural, urban, and suburban adult Greek populations. Hormones 11(4):458
4. Stewart B (2014) Wild, world cancer report. WHO press (e-PUB)
5. Mammas CS, Mandellos G, Economou GP, Lymberopoulos D (2001) Structuring expert-leaded medical protocols for telemedicine systems. In: Proceedings of the 23rd annual international conference of the IEEE engineering in medicine and biology society, IEEE procceedings, On Procceedings, IEEE, Constantinople, p 3529
6. Pickwell KM, Siersma V, Kars M, Holstein P, Schaper NC (2013) Eurodiale consortium. Diabetic foot disease: impact of ulcer location on ulcer healing. Diabetes Metab Res Rev 29:377
7. Mammas CS, Saatsakis G, Lemonidou C, Mamma AS, Chasiakos D (2016) Ergonomics of tele-cytology for remote pap-smear evaluation integrated with Big Data analytics and computing to optimize prevention of cervical cancer in developing countries. In: Proceedings of the 48th annual conference of nordic ergonomics and human factors society. NES 2016, p 225
8. Kainotomeis.gr (2104) Prometheus I (pn 1008239) a digital medical device for remote and personalized prevention of diabetes related complications-with an emphasis on diabetic foot. In: Proceedings of the third panhellenic competition of innovation-Greece innovates organized by the Hellenic industrial association and the Eurobank. Greece Innovates 2014, p 36. http://www.sev.org.gr/uploads/Documents/lefkoma.pdf. Last accessed 2019/08/01
9. Siersma VD et al (2013) Importance of factors determining the low health-related quality of life in people presenting with a diabetic foot ulcer: the eurodiale study. Diabet Med 30:1382
10. Karavatselou E, Economou GP, Chassomeris C, Daneli V, Lymperopoulos D (2001) OTE-TS-A new value-added telematics service for telemedicine applications. IEEE Trans Inf Technol Biomed 5:210
11. Coiera E (2015) Guide to health informatics. CRC, United States
12. Allen TC (2014) Digital pathology and federalism. Arch Pathol Lab Med 138(2):162
13. Fónyad L et al (2012) Validation of diagnostic accuracy using digital slides in routine histopathology. Diagn Pathol 7:35
14. Mammas CS. The biomedical technology "Prometheus" (pn 2003016 and pn 1008239) and personalized medicine in the field of transplantation and in the primary, secondary and tertiary prevention. In: Proceedings of the Bousssias conference on personalized medicine (BOUSSIAS 2013). http://personalizedmedicine.boussiasconferences.gr/default.asp?pid=23&la=1&pwID=268&remind=6. Last accessed 2019/08/01
15. Mammas CS. Prometheus (pn 2003016) and prometheus I (pn 1008239) patented digital medical devices. 1st Greek innovation forum-under the auspice of the Hellenic ministry of education (GIF 2014). http://www.greekinnovationforum.eu/p/1st-gif.html. Last accessed 2019/08/01
16. Mammas CS (2015) The remote evaluation of pap-test applying tele-cytology in the context of uterus graft remote evaluation and in the remote prevention of cervical cancer. In: Proceedings of the program of excellence 2014–16-Siemens program for Greece, vol IV. http://www.lib.uoa.gr/fileadmin/user_upload/I.K.Y_Papanikolaou.pdf. Last accessed 2019/08/01
17. Siersma V et al (2014) Health-related quality of life predicts major amputation and death, but not healing, in people with diabetes presenting with foot ulcers the eurodiale study. Diab Care 37:694

18. Panahiazar M, Taslimitehrani V, Jadhav A, Pathak J (2014) Empowering personalized medicine with big data and semantic web technology: promises, challenges, and use cases. In: Proceedings of the IEEE international conference on big data, p 790
19. Mammas CS (2017) Innovating training networks for interactive e-learning as the method of choice to benefit from big data analytics and computing in the remote multidisciplinary approach and decision making to optimize quality in cancer treatment. ETP722, Sages. https://www.sages.org/wp-content/uploads/2013/09/SAGES-2017-Final-Program.pdf. Last accessed 2019/08/01

Portuguese Firefighters' Anthropometrics: Pilot Study Results

Anna S. P. Moraes, Miguel A. F. Carvalho, Rachel S. Boldt, Fernando B. N. Ferreira, Susan P. Ashdown, and Linsey Griffin

Abstract Personal protective equipment is of paramount importance in aiding and protecting firefighters against hazardous conditions. Personal protective equipment should be compatible to firefighters' body dimensions. However, the lack of updated anthropometric data for this specific occupational group can lead to ill-fitting equipment, compromising work performance and the level of protection. To understand if Portuguese firefighters' protective equipment is adjusted to their anthropometrics a study designated as *Size FF Portugal* is currently underway. A pilot study was conducted in a fire brigade located in the North of Portugal aiming to collect preliminary anthropometric data. Thirty-two male firefighters participated in the pilot study. This paper contrasts some preliminary anthropometric data of Portuguese firefighters with an anthropometric database of U.S. firefighters. The results show statistically significant differences between the stature and weight of Portuguese and U.S. firefighters. Conversely, crotch height and calf circumference measurements were not statistically different when comparing firefighters of both countries. Final considerations regarding the study limitations are presented.

Keywords Anthropometric data · Firefighting · Ill-fitting · Personal protective equipment · Turnout gear

---

A. S. P. Moraes (✉) · M. A. F. Carvalho · R. S. Boldt · F. B. N. Ferreira
University of Minho, Guimaraes, Portugal
e-mail: sophiapiacenza@gmail.com

S. P. Ashdown
Cornell University, Ithaca, NY, USA

L. Griffin
University of Minnesota, Minneapolis, MN, USA

D. Sumpor et al. (eds.), *Proceedings of the 8th International Ergonomics Conference*, Advances in Intelligent Systems and Computing 1313,
https://doi.org/10.1007/978-3-030-66937-9_17

1 Introduction

Personal protective equipment is of paramount importance in aiding and protecting firefighters against multiple hazards confronted on a daily basis. It is thus critical that personal protective equipment (PPE) allows firefighters to perform their duties with minimal limitations and maximum safety and comfort [1]. Firefighting PPE should meet specific standards, be safely designed and constructed, as well as be adjusted to firefighters' body dimensions.

Despite efforts in improving firefighting equipment, the impacts of wearing PPE have been highlighted in a fair amount of research [2–6]. Fit and size of PPE remain common complaints among firefighters [1, 7–10]. Under extreme circumstances, garment dimensions can become critically determinants of the safety and performance of the design [11]. In this sense, access to reliable anthropometric data is paramount.

Anthropometric data must accurately represent the dimensions of those for which the design is intended [12]. However, the scarcity of reliable and updated anthropometric data on target-user populations can lead to an unsuccessful design [13].

This paper presents the preliminary results of an anthropometric study of Portuguese firefighters and compares these results with a database of U.S. firefighters. The diverse anthropometric measurements found for firefighters from each country are discussed. Moreover, a short review on Anthropometrics as well as a brief description of the study *Size FF Portugal* are presented as follows.

1.1 Anthropometric Studies

Anthropometry is essential for applying ergonomic principles to the design and improvement of a wide range of products for both specific users and the general population. Moreover, it is critical for the design of workspaces and tools. Hsiao [14] mentions that the importance of anthropometric data for product efficacy and safety has been recognized by various industries.

Anthropometric data have found one of their earliest commercial applications in sizing of clothing [11]. Cichocka et al. [15] emphasize that the design of a garment has to be preceded by knowledge of the human body morphology, with the characteristic features and proportions. Additionally, it is worth mentioning that specific population groups, including specialized professionals, may present different body dimensions. Hsiao et al. [16] used a U.S. anthropometric database to compare measurements of occupational subgroups. The authors identified that both and men and women serving in protective service occupations (including firefighting and fire prevention occupations) had the tallest mean standing heights and greater mean weights when compared to all other occupations included in the study [16].

1.2 *Firefighters' Anthropometric Surveys*

Despite the efforts of researchers and industries in conducting anthropometric surveys, databases of firefighters are still very limited. One of the first surveys exclusively with firefighters was in New Zealand, and focused on developing a sizing system for protective clothing and for purchases of other selected equipment [17].

Some years later, The National Institute for Occupational Safety and Health (NIOSH) led an anthropometric survey with the U.S. firefighting population. The study, conducted during 2009–2012, aimed to help improve the ergonomic and safety specification for fire apparatus, including PPE [18–20]. Another anthropometric study conducted in the U.S. was published in 2015 by Boorady [21]. The objective of this study was to assess the sizing of turnout ensembles worn by firefighters. However, this study involved a smaller sample size and all participants were from the Midwest region [21]. In Europe, the only anthropometric survey available reported data of 61 measurements collected from a sample of 316 British female firefighters [22].

1.3 *The Study* Size FF Portugal

Aiming to fulfill the lack of anthropometric data of firefighting population as well as to understand if Portuguese firefighters' protective equipment is adjusted to their anthropometrics, a study designated as *Size FF Portugal—Anthropometric Study of Portuguese Firefighters* is currently underway. Anthropometric data is being collected in fire brigades in Portugal. This paper presents the preliminary results of a pilot study conducted in a fire brigade located in the North of Portugal.

2 Materials and Methods

2.1 *Data Collection*

In order to conduct the data collection, an authorization from the chief-in-command was obtained. Data were collected during different working shifts and all firefighters were invited to participate. Participants' anthropometric data were collected through both direct and indirect techniques. Indirect measurements were obtained using a handheld 3D body scanner. Additionally, weight, stature, crotch height, calf circumference, and calf height measurements were directly manually-collected.

Weight, stature, and crotch height were acquired with participants barefoot and wearing underwear. The weight was obtained using a digital scale. Values were recorded to the nearest 100 g. The stature and the crotch height were measured using a moveable stadiometer. Values were recorded to the nearest 0.1 cm.

Measurements of the calf circumference and calf height were taken in the leg correspondent to the predominant foot with participants barefoot. The maximum calf circumference was recorded to the nearest 0.01 in. The calf height was measured on the outside of the calf, from the floor up to the fullest part of the calf, using a retractable tape measure.

All measurements were collected by the same researcher to avoid intermeasurer differences. Measurements were annotated in a proper worksheet for further analysis.

2.2 Data Analysis

Subsequently data were treated and analyzed. A screening was made to ensure that there were no outliers. Measurements recorded in inches were converted to the metric system. Descriptive statistics were applied to analyze the collected measurements. Inferential statistics were used to compare collected data and the U.S. database.

3 Results and Discussion

Thirty-two male firefighters agreed to participate in the pilot study and completed all data collection stages.[1] In this paper, anthropometric data that were directly collected, i.e. the stature, crotch height, weight, calf circumference and calf height are presented.

3.1 Demographic Information

From the 32 participants, 24 were both career and volunteer firefighters and 7 were volunteer firefighters. The assistant chief responsible for the brigade also participated in the study. The average age of participants was 38.8. Participants' years of experience in firefighting were 17.7 on average, ranging from 2 months to 38 years.

3.2 Descriptive Statistics Analysis

Summary statistical analysis on the 5 anthropometric data were performed for the arithmetic mean, standard deviation, range, and percentiles. Such data is presented in Table 1.

[1]Some participants decided to interrupt their participation and others were not available to conclude all data collection stages.

Table 1 Descriptive statistics of Portuguese anthropometric data

	Stature (mm)	Crotch height (mm)	Calf circumf (mm)	Calf height (mm)	Weight (kg)
Sample size	32	32	32	31	32
Mean	1723.5	776.7	394	352.7	83.7
STD	48.9	35.8	33.8	17.7	13.8
Min–Max	1610–1812	715–845	330.2–449.6	310–385	58–112.6
5th percentile	1642.7	715	340.6	326.5	63.1
50th percentile	1728	780	393.7	353	84.1
95th percentile	1788.5	822.3	448.2	380	107.3

Results show that Portuguese firefighters' participants in the pilot study were on average 1723.5 mm tall. As expected, stature measurements presented the highest standard deviation. The crotch height mean was 776.7 mm and the standard deviation 35.8 mm. Calf circumference and calf height presented lower values of standard deviations, which were also expectable. The average calf circumference was 32 mm and the average calf height was 352.7 mm. The mean weight of participants was 83.7 kg (SE mean = 2.4, STD = 13.8), ranging from 58 to 112.6 kg.

3.3 Anthropometric Database of U.S. Firefighters

As mentioned before, the first comprehensive anthropometric database of U.S. firefighter population was released by The National Institute for Occupational Safety and Health [18, 20]. The survey sampling plan included firefighters from different U.S. states, grouped in 4 regions.[2] The final sample included 863 male and 88 female firefighters. Seventy-one body dimensions relevant to the design of different fire equipment were measured using diverse techniques and tools.[3] Anthropometric data relevant to this study are presented in Table 2.

3.4 Comparison of Anthropometric Data of Both Countries

The anthropometric data of Portuguese firefighters obtained in the pilot study were compared to the NIOSH Firefighters Anthropometrics database. The calf height,

[2]Pacific West, North Central, Northeast, and South.

[3]The complete database as well as a description of data collection procedures are available in https://www.cdc.gov/niosh/data/datasets/rd-1007-2015-0/default.html.

Table 2 NIOSH U.S. firefighters anthropometric data

	Stature (mm)	Crotch height (mm)	Calf circumf. (mm)	Weight (kg)
Sample size	863	863	863	863
Mean	1769	785.4	398.1	93
STD	66.7	43.7	29.2	14.8
5th percentile	1659.7	713.3	353	71.3
50th percentile	1766.8	785.9	396	91.4
95th percentile	1881.3	858.4	449	120.4

Source NIOSH [20]

despite being relevant to the design of fire boots, was not included in the U.S. survey and could not be compared. All dimensions of Portuguese firefighters presented high values of standard error, which can be explained by the small sample size.

The Student's t-test for independent means (unpaired test) was applied for all other considered measurements. The two-tailed tests were performed assuming a normality distribution of data. The null hypothesis proposed no difference between the means of each measurement. In addition, the F-test of equality of variances was applied for assessing whether firefighting populations variances of each considered measurement should be considered equal or not. According to the F-test for variance, the sample standard deviation of stature of Portuguese firefighters was not equal to the sample standard deviation of stature of U.S. firefighters, which means that the difference between the sample standard deviation of both populations' stature was big enough to be statistically different. For all other considered dimensions (crotch height, calf circumference, and weight) the results of the F-test showed that sample standard deviations of both populations can be considered to be equal at a significance level of 95%. From the results of the F-tests, appropriate Student's t-tests were applied (considering equal or unequal variances), taking into account the unequal sample sizes. The statistic t, critical t values (t_c), degrees of freedom, and p-values are presented in Table 3.

Concerning the stature, the results of the t-Student test revealed that the null hypothesis is rejected at a significance level of 0.05. It is possible to conclude that there is a statistically significant difference between the stature of Portuguese firefighters and U.S. firefighters. The null hypothesis of the t-Student test for the crotch height was not rejected for the same significance level, meaning that there is not enough evidence to assume that the crotch height of the population of Portuguese firefighters is different than the U.S. firefighters. Similarly, differences between calf circumferences of both firefighters' populations are not statistically significant (significance level 95%). However, differences between firefighters' weight can be considered to be extremely statistically significant at the same significance level.

Table 3 T-Students' tests of considered measurements

	Stature (mm)	Crotch height (mm)	Calf circumf. (mm)	Weight (kg)
t	−5.091	−1.112	−0.775	−3.498
t_c	2.029	1.963	1.963	1.962
p-value (two tailed)	0.0000	0.663	0.4383	0.0005
Degrees of freedom	35.419	893	893	893

Different anthropometric characteristics of specialized occupational groups justifies the need of a better understanding of firefighters' body dimensions. Despite preliminary, the presented results evidence the fact that Portuguese and U.S. firefighters present different anthropometrics. Such differences are relevant and must be considered by manufacturers in the design of firefighting PPE.

4 Conclusion and Final Considerations

The lack of available and updated anthropometric data for specific occupational groups can lead to ill-fitting equipment. In the case of firefighters, it can compromise protection and performance. Many authors have pointed out the need for a better understanding of firefighters' body dimensions to address fit and sizing issues in PPE.

The study *Size FF Portugal* aims to provide a comprehensive anthropometric database of Portuguese firefighters to assist the design of firefighting PPE. This paper presented the preliminary data of a pilot study conducted in a fire brigade located in the North of Portugal. Five anthropometric measurements are presented and 4 are compared to a U.S. firefighting database. Results show significant differences between the stature and weight of Portuguese and U.S. firefighters. On the contrary, calf circumference and crotch height measurements are not statistically different when comparing both firefighting populations. Although the reduced sample size of the Portuguese firefighting data, the preliminary results indicate that specific anthropometric databases are needed, aiming to accommodate as many of the target-user population as possible.

Small differences in terms of data collection procedures adopted in the NIOSH survey and the *Size FF Portugal* pilot study were observed, which deserve a better investigation in order to ensure that data are reliable and comparable.

Acknowledgements We would like to acknowledge the 2C2T-Centre for Textile Science and Technology of the University of Minho. This work is financed by FEDER funds through the Competitive Factors Operational Program (COMPETE) POCI-01-0145-FEDER-007136, by national funds through the FCT-Portuguese Foundation for Science and Technology under the project UID/CTM/000264, by Fundo de Apoio às Vítimas dos Incêndios de Pedrógão, and by ICC/Lavoro.

References

1. Boorady LM, Barker J, Lee YA, Lin SH, Cho E, Ashdown SP (2013) Exploration of firefighter turnout gear Part 1: identifying male firefighter user needs. J Text Apparel Technol Manag 8:1–13
2. Coca A, Williams WJ, Roberge RJ, Powell JB (2010) Effects of fire fighter protective ensembles on mobility and performance. Appl Ergon 41:636–641. https://doi.org/10.1016/j.apergo.2010.01.001
3. Son S-Y, Xia Y, Tochihara Y (2010) Evaluation of the effects of various clothing conditions on firefighter mobility and the validity of those measurements made. J Human-Environ Syst 13:15–24. https://doi.org/10.1618/jhes.13.15
4. Park H, Kim S, Morris K, Moukperian M, Moon Y, Stull J (2015) Effect of firefighters' personal protective equipment on gait. Appl Ergon 48:42–48. https://doi.org/10.1016/j.apergo.2014.11.001
5. Park H, Trejo H, Miles M, Bauer A, Kim S, Stull J (2015) Impact of firefighter gear on lower body range of motion. Int J Cloth Sci Technol 27:315–334. https://doi.org/10.1108/IJCST-01-2014-0011
6. Angelini MJ, Kesler RM, Petrucci MN, Rosengren KS, Horn GP, Hsiao-Wecksler ET (2018) Effects of simulated firefighting and asymmetric load carriage on firefighter obstacle crossing performance. Appl Ergon 70:59–67. https://doi.org/10.1016/j.apergo.2018.02.006
7. Huang D, Yang H, Qi Z, Xu L, Cheng X, Li L, Zhang H (2012) Questionnaire on firefighters' protective clothing in China. Fire Technol 48:255–268. https://doi.org/10.1007/s10694-011-0214-0
8. Park H, Hahn KHY (2014) Perception of firefighters turnout ensemble and level of satisfaction by body movement. Int J Fash Des Technol Educ 7:85–95. https://doi.org/10.1080/17543266.2014.889763
9. Park H, Park J, Lin SH, Boorady LM (2014) Assessment of firefighters' needs for personal protective equipment. Fash Text 1:1–13. https://doi.org/10.1186/s40691-014-0008-3
10. Lee J-Y, Park J, Park H, Coca A, Kim JH, Taylor NAS, Son S-Y, Tochihara Y (2015) What do firefighters desire from the next generation of personal protective equipment? Outcomes from an international survey. Ind Health 53:434–444. https://doi.org/10.2486/indhealth.2015-0033
11. Gupta D (2014) Anthropometry and the design and production of apparel: an overview. In Gupta D, Zakaria N (eds) Anthropometry, apparel sizing and design. pp 34–66. Woodhead Publishing, Cambridge. https://doi.org/10.1533/9780857096890.1.34
12. Barroso MP, Arezes PM, Da Costa LG, Miguel AS (2005) Anthropometric study of Portuguese workers. Int J Ind Ergon 35:401–410. https://doi.org/10.1016/j.ergon.2004.10.005
13. Dianat I, Molenbroek J, Castellucci HI (2018) A review of the methodology and applications of anthropometry in ergonomics and product design. Ergonomics 61:1696–1720. https://doi.org/10.1080/00140139.2018.1502817
14. Hsiao H (2013) Anthropometric procedures for protective equipment sizing and design. Hum Factors 55:6–35. https://doi.org/10.1177/0018720812465640
15. Cichocka A, Bruniaux P, Frydrych I (2014) 3D garment modelling—creation of a virtual mannequin of the human body. Fibres Text East Eur 22:123–131
16. Hsiao H, Long D, Snyder K (2002) Anthropometric differences among occupational groups. Ergonomics 45:136–152. https://doi.org/10.1080/00140130110115372
17. Laing RM, Holland EJ, Wilson CA, Niven BE (1999) Development of sizing systems for protective clothing for the adult male. Ergonomics 42:1249–1257. https://doi.org/10.1080/001401399184929
18. Hsiao H, Whitestone J, Kau TY, Whisler R, Routley JG, Wilbur M (2014) Sizing firefighters: method and implications. Hum Factors 56:873–910. https://doi.org/10.1177/0018720813516359

19. Hsiao H, Weaver D, Hsiao J, Whitestone J, Kau TY, Whisler R, Ferri R (2014) Comparison of measured and self-reported anthropometric information among firefighters: implications and applications. Ergonomics 57:1886–1897. https://doi.org/10.1080/00140139.2014.952351
20. NIOSH, National Institute for Occupational Safety and Health (2015) Firefighter dataset—NIOSH RD-1007-2015-0. https://www.cdc.gov/niosh/data/datasets/rd-1007-2015-0/default.html
21. Boorady LM (2015) Bunker gear for fire fighters: does it fit today's fire fighters? J Text Apparel Technol Manag 9:1–15
22. Stirling M, National anthropometry survey of female firefighters—designing for safety, performance and comfort. Tamworth. https://www.humanics-es.com/FireFighterAnthropometry.pdf

Embedding Function Icons into QR Codes

Filip Cvitić, Jesenka Pibernik, Josip Bota, and Ante Poljičak

Abstract This paper proposes an approach to produce high-quality visual QR codes introducing the concept of function icons (FI)—images embedded into QR code. FI design resembled abstract symbols, like identifiable and memorable smartphone app icons or logos. The embedding visually points to the function of a QR code and therefore is more attractive to the end-user than the traditional QR code. Also, the FI perceptual simplicity increases users' navigation and accessibility over different media. For optimization, the proposed FI design was altered by pixel size, color, and luminance modification, without affecting the decoding reliably. The contrast between dark and light modules of the FI color tone was determined to a minimum of 70% to enable the code reader's conversion of the image to grayscale and then segmentation it into black and white pixels. The eye-tracking research on QR codes with implemented icons confirmed the subjective attractiveness evaluation. The technical quality of the proposed design was verified with a QR code readability test. Since more than 95% of examinees successfully recognized the content behind a QR code before scanning and evaluated the design as more attractive than a traditional solution, we demonstrated that our approach produces high-quality results for a variety of FI designs.

Keywords Function icons · Contrast · QR code · Readability · Accessibility

1 Introduction

QR codes are visually unreadable and the end-user cannot identify the purpose of the QR code without the aid of a bar code reader [1]. Before incorporating visual features into a QR code it is essential to ensure the technical qualities—i.e., that the bar code can be reliably, accurately, and correctly encoded and decoded, yielding all the information contained without error. There have been several efforts to

F. Cvitić · J. Pibernik (✉) · J. Bota · A. Poljičak
University of Zagreb Faculty of Graphic Arts, 10 000 Zagreb, Croatia
e-mail: jpiberni@grf.hr

D. Sumpor et al. (eds.), *Proceedings of the 8th International Ergonomics Conference*, Advances in Intelligent Systems and Computing 1313,
https://doi.org/10.1007/978-3-030-66937-9_18

improve the appearance of such embeddings [2–5] which can be classified into two categories, methods that modify the luminance or color of image pixels and methods that replace QR modules [6, 7]. This paper is exploring the first category, attempting to introduce the image icon based on the modification of the pixel's size, color, and luminance, without affecting the decoding robustness. For the research, an introduced image embedded into the QR code is named function icon (FI). FI is a proprietary design, inspired by web icons, developed through earlier research [8] to maximize visibility and recognizability of the QR function. Several FI designs, based on icons for popular apps (YouTube, Facebook, Skype, Instagram, Geolocation, Twitter) have been developed and embedded into matching QR codes (Fig. 1).

The novelty of this work is based on quality of user experience design approach. The proposed manipulations of QR code modules could be done by a graphic designer without the use of any algorithms. The goal was to optimize QR code design for print so that the printed image of the FI would be sufficiently visible and the successful reading of the QR code at both high speed and accuracy preserved. We assumed that QR codes with FI would keep their readability while enhancing user satisfaction, accessibility, attractiveness, and comprehension of the informational content.

1.1 Technical Limitations of QR Codes

The successful reading of a bar code also depends, among other things, on the camera resolution and on the pixels per module of the QR code itself. In most cases, QR code images taken by a mobile phone camera are coded in the RGB color space for screen reading. In order to decrease the cost of information storage, the bar code scanning application converts the image to grayscale, and then segments it into black and white pixels [9, 10]. This is followed by decoding the bitstreams. The methods of decoding the image rely upon image denoising [11] or on-camera shaking. Based on the number of different decoding algorithms to read QR codes

Fig. 1 Samples of QR codes with different FI, different colors, and different module sizes: FI are the following: **a** magenta Twitter, **b** magenta Facebook, and **c** magenta Instagram

[12–14] it is noticeable that no single standardization method of reading a bar code exists. The principal technical challenge to any method of embedding an image into the QR code is the possibility of decoding it with standard applications [15, 16].

1.2 Accessibility of the QR Code

Accessing content is the main reason why users scan QR codes. Therefore, together with technical requirements, user's need to quickly and trustworthy visually recognize which type of content they are going to receive is of utmost importance. QR codes are being used to send audiences to a website–quicker than typing the link, to bookmark a web page, to initiate phone calls, send short messages, send emails, produce links to web URL's, connect to WI-FI networks, access information, get coupons, view videos, purchase items, process orders, advertise products, etc. Those purposes are mainly identical to the most common applications for smartphones or computers. In computing, hyperlink, or file shortcut to access the program or data is commonly represented by icons. The icon itself is a quickly comprehensible graphical user interface element, accessible on the system and is more like a traffic sign than a detailed illustration of the actual entity it represents.

2 The Experiment

The experiment in this paper consisted of three steps. The first step was surveying smartphone users to determine influential factors for sample preparation. The second one was–subjective attractiveness evaluation and eye-tracking objective research. The eye-tracking experiment was to prove that number of eye gazes is correlated positively to the degree of attractiveness. The third one–the QR code readability testing. Finally, the results of subjective and objective methods data were statistically interpreted. The following is the explanation of each part.

2.1 Online Questionnaire

The online survey has been distributed by using the Survey gizmo service (http://www.surveygizmo.com/s3/3717594/QR-design). The investigation form was completed by 62 smartphone users. The users have been selected by convenience sampling method. 62.5% of participants fit into the category of age 16–29; 26.6% fit into the category of age 30–44, and 10.9 fit into the category of age above 45. 35. 9% of participants were male, while 64.1% were female. Regarding the educational level, 43.8% of participants reported having a faculty degree, 42.2% reported as having a Master of Science degree, and 12.5% as holding a Ph.D.

The results have verified the relevant conclusions from previous research [3] that there are many identical purposes for which users scan QR codes and use smartphones, most frequently: accessing online maps and navigation, accessing WiFi, obtaining discounts, doing business transactions, listen to music, browsing the Internet, online shopping, receiving rewards, etc. Participants were 95% correct in assuming which information is encoded in the QR code with embedded FI. Also, the study identified the most influential factors according to examined users' satisfaction in the context of QR scanning. Those factors were: quality of information, achieving the desired action, and procedure for accessing. To adopt QR scanning services factors such as perceived usefulness, perceived ease of use, and perceived risk must be considered. Besides, the result analysis of participants' aesthetic preferences of QR code design has determined the selection of factors to be manipulated in the experimental phase.

2.2 *QR Code Sample Preparation*

QR codes were generated using Kerem Erkan's QR code generator. For the needs of this experiment, the FI has been embedded into version 3 (29 × 29 modules) with the L redundancy level. FI images were positioned over the QR codes and segmented using the pathfinder tool in Adobe Illustrator. To maintain high reliability, the area coverage of FI was determined as 20% (max limit) of the total QR code area.

The FI was created as vector images to be more attractive to the users. The central modules in the QR codes were resized from 1.43 to 0.48 mm to increase the perceptual recognizability of the FI. To facilitate reading, the brightness and pixel colors in CMYK of the embedded image were manipulated, as well as the size of certain modules of the QR code [17, 18].

Module center spacing remained constant. Considering that, ideally, only the pixels in the center of the QR code are relevant for correct decoding the resized modules were modeled based on data obtained in the article [19]. The light and dark modules of FI were prepared by applying two different percentages of cyan and magenta color coverage each (from 100/80% to 40/10%). The Symbol Contrast grade (SC) was determined as 4,0 (A) if SC_70%, 3.0 (B) if SC_55%, 2.0 (C) if SC_40%, 1.0 (D) if SC_20%. Due to the high lightness properties of yellow color, the darker tones were achieved by adding 10, 20, and 30% key black. The samples were printed in accordance to Print Quality Guidelines given in ISO/IEC 18004:2000(E) [20] on 230 g/cm^3 Premium Photo Paper (format A4), from PRINT-RITE, printed with Epson stylus pro 3800 using color management control (EFI color proof xf) to achieve the defined ink coverage.

2.3 QR Code Attractiveness Evaluation

The study was conducted at the User Experience Lab at the Faculty of Graphic Arts, University of Zagreb, in a lightly obscured area of constant light conditions of neutral coloration. The research was done with eye-tracking equipment (Tobii X60) and monitor 22-inch Lenovo ThinkVision L2251x on a resolution of 1680 × 1050 pixels (with the ratio 16:10). A total of 31 examinees participated (7 male and 24 female participants), all of which had an unlimited amount of time available for the research. From the total number of examinees, 26 were students (83.87%), four (12.9%) had a Master of Science degree and one (3.23%) had a Ph.D. 27 (87.1%) were in the age group of 16–29 and 4 (12.9%) of them were in the age group of 30–44. Each examinee was positioned according to the required conditions defined in the device's instructions or 50 cm from the monitor. Each examinee was given instructions regarding the task.

Before the eye-tracking research, the task was explained to the examinee and he/she had to complete an anonymous questionnaire like the survey in experiment 1. After the calibration of the equipment, the examinee was presented with the onscreen display of a traditional QR code, not included in the research. QR code research samples were placed on a white background and divided into two fractions: group and individual samples. The order of slides was randomly changed for each examinee. Each QR code displayed to the examinee differed in color (cyan—C, magenta—M or yellow—Y), and in function icon design implemented in the QR code (Facebook—F, Skype—S, Twitter—T, YouTube—Y, Instagram—I and Geolocation—G). The first letter in the following tables represents color, the second one FI, and the third one module size. In the beginning, six groups of six QR code design variations (with one function icon design) were shown to the examinee. In each group view, the examinee had to choose the most attractive QR code with a mouse click.

After displaying all the function icon design groups, the examinee had to evaluate 36 QR code samples individually. Followed by each sample, the examinee was asked "How would you rate the design quality of the displayed QR code?": (1) "Design is not attractive", (2) "Design is attractive", (3) "Design is very attractive" (Table 1). The Sum of total gazes for each FI design displayed individually is shown in Table 2. The most attractive individual FI is shown in Fig. 2.

2.4 QR Code Scanning

Different variations of color contrast (ink coverage) of the FI's in the QR codes samples were scanned using two smartphones (Lenovo ZUK Z2, iPhone 6) with applications seen in Table 3. The applications where used in their default settings. When selecting the QR code scanning applications for two criteria were used: number of downloads and user rating.

Table 1 The attractiveness evaluation for each FI

YFS	MFL	MFS	CFS	YFL	CFL	MYL	MYS	YYL	YYS	CYS	CYL	CGL	MGS	YGL	YGS	CGS	MGL
77	67	65	70	67	73	68	73	64	76	63	65	59	55	50	59	65	63
CTS	**YTS**	**YTL**	**MTS**	**CTL**	**MTL**	**MSL**	**MSS**	**CSL**	**YSL**	**CSS**	**YSS**	**YIS**	**CIL**	**YIL**	**MIL**	**MIS**	**CIS**
52	58	65	60	66	65	63	65	63	54	54	59	59	66	65	65	75	63

Table 2 Sum of all gazes for each FI (in milliseconds)

CFL	CFS	YFL	MFL	CSL	CTL	CIL	CGL	CGS	MGL	MGS	YGL
352560	56644	82994	70887	73949	69436	63425	69326	61510	66320	59810	57730
YYS	**CSS**	**CTS**	**CIS**	**MTS**	**MIS**	**MSS**	**MFS**	**YFS**	**YSS**	**YTS**	**YIS**
47354	61546	53399	45440	52387	55212	69427	49852	54333	53917	49203	58925
YGS	**CYL**	**CYS**	**MYS**	**MYL**	**YYL**						
47355	61777	44259	49216	50551	44271						

Fig. 2 The most attractive QR codes selected individually: **a** magenta Instagram, **b** cyan Geo location, and **c** magenta YouTube

Table 3 Application specifications used in the research

Android (ZUK Z2)	iOS (Apple iPhone 5)
QR code reader (Scan) 1.2.2	Default camera app
QR barcode scanner Pro 1.2.70	i-nigma version 3.19.01
QR code reader (TWMobile) 2.6.1	QR reader version 1.7
Barcode scanner 4.7.7	

During scanning of the test sample constant light properties (620-1200 lx at 4150 K) were achieved using the X-RiteThe Judge II Viewing Booth under CWF for the cool white, fluorescent source settings. Each QR code was scanned within a 3 s timeframe by the device which was handheld by the researcher. To confirm that the QR code was read correctly by the application, each sample was scanned 3 times in a row and marked as «read» or «not read». Hand shaking, distance between the QR code and the smartphone's camera, scanning angle and phone technology (sensor, aperture) were not considered.

3 Results and Discussion

All statistical tests were two tailed using a 0.05 level of significance, unless otherwise noted. Pearson correlation analysis was performed to investigate the associations between the participants' evaluations of the QR codes and observation time. There was a positive relationship between the mean attraction ratings and mean rating observation time, $r = 0.413$, $p < 0.05$, indicating that longer-viewed QR codes were correlated with higher attraction ratings.

To examine whether QR code design features (i.e. color) affect the rating observation time, nonparametric tests were used since the Kolmogorov-Smirnov test of normality showed the assumption of normality was violated ($p < 0.001$ for all cases).The Friedman test revealed that there were significant differences in rating observation time between different colored QR codes $\chi 2(2) = 33.87$, $p < 0.001$. Follow-up Wilcoxon tests were used with Bonferroni correction, so only the

p-values below 0.017 were considered significant. Cyan QR codes were viewed longer ($M = 2762.18$, SD = 3346.85) than magenta QR codes ($M = 1802.69$, SD = 1205.15), $z = -4.99$, $p < 0.001$, and yellow QR codes ($M = 1832.06$, SD = 1363.65), $z = -5.85$, $p < 0.001$. However, there was no significant difference in rating observation time between yellow and cyan QR codes, $z = -0.14$, $p > 0.05$. Pearson correlation analysis of QR code readability results revealed no significant association between the mean attraction ratings and mean QR code readability, $p = 0.355$.

Because the focus of this paper was user's attractiveness of QR codes with implemented FI the restrictions of readability test for this experiment were the following. Dark and light modules in the FI of the QR codes were prepared with different ratios (from 100/80 to 40/10). The examined contrast for cyan and magenta was between 20, 30 and 40% of ink coverage. The experiment didn't involve higher contrast levels because the goal was to test low contrast levels and most users prefer to perceive the FI as a coherent line which makes the whole QR code more attractive. Better QR code readings in low contrast scenarios of the FI are found when dark modules are in the higher and mid-range color ink coverage (50–70%). Adding 30% of key black to yellow color mostly enables the correct readings of the QR codes.

4 Conclusion

It is possible to see that the most readable QR codes are not the most attractive ones. Based on this research results, we can conclude the user will always consider the most recognizable, the object that has the clearest outlines of the observed character within the QR code. That is, the one where the color saturation between the dark and light data modules is similar, approximately equal. For the proper decoding of the QR code the computer will need significant Symbol Contrast, minimum of 70% between the 0 value and the 1 value. Central module widths and heights shall span at least five image pixels. For readability purposes, it is important that the contrast is the highest possible, and from the user standpoint it is desirable for the implemented image to have the lowest possible contrast. Based on the previous conclusion and inside the experiments restrictions one can also see that the most attractive FI is the one with which they are most familiar from previous experience. Therefore, based on its simplicity of implementation, the high recognizability of the content, and the survey research which showed that examinees found QR codes with implemented FI more attractive than traditional QR codes, this method merits has been proven.

References

1. Zachi B., Ramakrishna K.: Visually significant QR codes: Image blending and statistical analysis," *IEEE International Conference on Multimedia and Expo (ICME).* (2013)
2. Lin Sh, Hu M, Lee Ch, Lee T (2015) Efficient QR code beautification with high quality visual content. IEEE Trans Multimedia 17(9):1515–1524
3. Skawattananon C, Vongpradhip S (2013) An improved method to embed larger image in QR code. In: The 2013 10th international joint conference on computer science and software engineering (JCSSE), Maha Sarakham, Thailand
4. Yu-Hsun L, Chang Y, Wu J (2013) Appearance-based QR code beautifier. IEEE Trans Multimedia 15(8):2198–2207
5. Yi-Shan L, Sheng-Jie L, Bing-Yu Ch (2013) Artistic QR code embellishment. Comput Graph Forum 32(7):137–146
6. Ming S, Zhenkun F, Longsheng F., Fan Z (2010) Identification of QR codes based on pattern recognition. In: World Automation Congress. Kobe, Japan
7. Garateguy GJ, Arce GR, Lau DL, Villarreal OP (2014) QR images: optimized image embedding in QR codes. IEEE Trans Image Process 23(7): 2842–53
8. Cvitić F, Pibernik J (2014) Decoding different patterns in various grey tones incorporated in the QR code. Acta Graphica 25(1–2):11–22
9. Kikuchi M, Fujiyoshi M, Kiya H (2013) A new color QR code forward compatible with the standard QR code decoder. In: International symposium on intelligent signal processing and communication systems. Naha, Japan
10. Muñoz-mejías D, González-díaz I, Díaz-de-maría F (2011) A low-complexity pre-processing system for restoring low-quality QR code images. IEEE Trans Consum Electron 57(3): 1320–1328
11. Lin CC, Chen MS (2009) A general scheme for QR-code image denoising on the camera phone. M.S. thesis, Department of Electrical Engineering, National Taiwan University
12. Yuji K, Daisuke D, Tomokazu T, Ichiro I, Hiroshi M (2011) Low resolution QR-code recognition by applying s-resolution using the property of QR-codes. In: International conference on document analysis and recognition—ICDAR '11. Beijing, China
13. Xiong Z, Cuiqun H, Guodong L, Zhijun L (2010) A binarization method of quick response code image. In: 2nd international conference on signal processing systems ICSPS. Dalian, China
14. Zhou J, Liu Y, Li P (2010) Research on binarization of QR code image. In: International conference on multimedia technology ICMT. Ningbo, China
15. Otsu N (1979) A threshold selection method from gray-level histograms. IEEE Trans Syst Man Cyber 9(a1):62–66
16. Chen Ch, Kot A, Yang H (2013) A two-stage quality measure for mobile phone captured 2D barcode images. Pattern Recogn 46(9):2588–2598
17. Ono S, Morinaga K, Nakayama S (2008) Two-dimensional barcode decoration based on real-coded genetic algorithm. In: IEEE Congress on Evolutionary Computation (IEEE World Congress on Computational Intelligence). Hong Kong, China
18. Samretwit D, Wakahara T (2011) Measurement of reading characteristics of multiplexed image in QR code. In: Third international conference on intelligent networking and collaborative systems (INCoS). Fukuoka, Japan
19. Hung-Kuo C, Chia-Sheng C, Ruen-Rone L, Niloy M (2013) Halftone QR codes. ACM Trans Graph 32(6):217:1–217:8, November
20. ISO, ISO/IEC 18004:2000 Information technology—Automatic identification and data capture techniques—Bar code symbology – QR Code. ISO/IEC, Geneva, 2000

Suitability of School Furniture and Correlation with Pain Prevalence Among Slovenian 6th to 9th Graders

N. Podrekar, K. Kastelic, M. Burnard, and Nejc Šarabon

Abstract School furniture has been recognised as a potential risk factor for musculoskeletal pain among primary school pupils. The aim of this study was to assess student-furniture mismatch and to determine the association between unsuitable school furniture and musculoskeletal pain among Slovenian pupils. Firstly, pupils' body dimensions were measured, and student-furniture mismatch was calculated. Secondly, pain prevalence was assessed using adjusted Nordic questionnaire. A total of 68 pupils from 6th to 9th grade (age 12 ± 1, 44% males) participated in the study. Seat height was unsuitable for 90% of pupils and desk height for 82% of pupils. One third of pupils reported knee pain and more than one fifth of pupils reported neck, shoulder, upper, and lower back pain. Logistic regression revealed no significant impact of school furniture on pain occurrence among pupils. However, the data indicated upper back pain might be associated with unsuitable backrest. The results showed that school furniture in Slovenian primary school is unsuitable for most pupils. The association between furniture and pupils' musculoskeletal pain needs further investigation. Redesign of school furniture is desired, and the anthropometric data obtained could be considered when designing new school furniture.

N. Podrekar (✉) · K. Kastelic · M. Burnard · N. Šarabon
InnoRenew CoE, Izola, Slovenia
e-mail: nastja.podrekar@innorenew.eu

K. Kastelic
e-mail: kaja.kastelic@innorenew.eu

M. Burnard
e-mail: mike.burnard@innorenew.eu

N. Šarabon
e-mail: nejc.sarabon@fvz.upr.si

N. Podrekar · N. Šarabon
Faculty of Health Sciences, University of Primorska, Izola, Slovenia

K. Kastelic · M. Burnard
Andrej Marušič Institute, University of Primorska, Koper, Slovenia

D. Sumpor et al. (eds.), *Proceedings of the 8th International Ergonomics Conference*, Advances in Intelligent Systems and Computing 1313,
https://doi.org/10.1007/978-3-030-66937-9_19

Keywords Ergonomics · Students · Student-furniture mismatch · Musculoskeletal pain

1 Introduction

Musculoskeletal pain is a global health problem, affecting individuals in all age groups and socio-demographic society [1]. The issues related with musculoskeletal pain are anticipated to increase with age [1]. Therefore, actions to prevent musculoskeletal conditions should target all age groups, including children. Moreover, it has been suggested that musculoskeletal conditions obtained during childhood may predispose individuals to musculoskeletal-related issues latter in adulthood [2, 3].

Internal factors, such as age and sex, have been recognised as a risk factor for musculoskeletal pain development in children [4, 5]. Among external factors, a lot of research has been conducted to investigate the impact of backpack carrying [6–8] on musculoskeletal pain. Indeed, backpack weight, total time wearing, and improper backpack wearing were identified to increase the risk for musculoskeletal pain among children [9, 10]. As children grow older, they spent more time seated in school which may also influence the development of musculoskeletal pain [11, 12].

School furniture, as one of the main micro-elements of the school system [13, 14], influences pupils' posture when seated in the classroom. Therefore, the dimensions of the school furniture should suit the anthropometric dimensions of the pupils to ensure comfortable sitting and encourage proper posture [15]. The discrepancy between the school furniture dimensions and pupils' body dimensions has been defined as the student-furniture mismatch. The studies conducted so far have shown a high student-furniture mismatch, ranging from 60 to 99% for seat height, from 55 to 99% for seat depth, and from 52 to 99% for table height [16–22]. Moreover, unsuitable school furniture has been associated with musculoskeletal pain among pupils. Particularly, a too low school desk has been shown to be a risk factor shoulder and neck pain [23, 24], and a too low backrest for lower back pain [23].

The suitability of school furniture in Slovenian schools is lacking and the knowledge of pain prevalence among Slovenian pupils is scarce. The aim of this study was to explore whether the mismatch between the school furniture and body dimensions of pupils, impacts musculoskeletal pain among Slovenian primary school pupils. We hypothesised that higher mismatch will result in increased musculoskeletal pain in certain body regions.

2 Methods

Pupils from 6th to 9th grade of primary school participated in this cross-sectional study. Before the beginning of the study, the informed consent, signed by pupils and their parents, was obtained. Pupils who volunteered were generally healthy and had no physical or mental disabilities. The study was approved by the Slovenian National Medical Ethics Committee (number of approval 0120-631/2017/2). The data analysed in the study were completely anonymous.

2.1 Anthropometric Measurements

The anthropometric measurements were performed manually using GPM anthropometric measurement set (Gneupel Präzisions-Mechanik, Swizterland). During the measurement the pupils wore light clothes and no shoes. They sat upright on a chair with adjustable seat height without a backrest having ankles, knees and hips bent at 90°, and feet completely on the floor. The anthropometric measures included: shoulder height, subscapular height, elbow height, popliteal height, thigh thickness, hip width and buttock-popliteal length, body height, body mass. The anthopometric measurements were performed as suggested by Castellucci et al. [25]. Each measurement was taken three times and the average values were considered for further analyses.

2.2 School Furniture Measurements

The classroom furniture was measured with the GPM tape and the electronic inclinometer (Baseline Evaluation Instruments, Fabrication Enterprise Inc., USA). The furniture measures included: seat height, seat depth, seat width, upper edge of backrest, seat to desk clearance, desk height, and seat inclination. The measurements were performed as suggested by Castellucci et al. [25]. Measurements were taken two times and the average values were used for further analyses.

2.3 Student-Furniture Mismatch Calculation and Pain Assessment

The student-mismatch was calculated using the equations proposed by Castellucci et al. [22]. Because the pupils wore slippers at school, the shoe correction of 0.5 cm was used. Pain prevalence was assessed using adjusted Nordic questionnaire [23]. The questionnaire consisted of a manikin, showing eight body regions and nine questions related to pain.

2.4 *Statistical Analysis*

The data obtained were transcribed into the Microsoft Excel (Microsoft 365, 2020) and analysed with R 3.6.1 (R Core Team (2019) using R Studio 1.2.1335). A logistic regression was used to assess the relationship between one dependent and a set of independent variables. The statistical significance threshold was set at $p < 0.05$.

3 Results

A total of 68 pupils participated in the study. Detailed participants characteristics are presented in Table 1 and body dimensions of pupils are displayed in Table 2.

The highest mismatch was recognised for seat (89.1%) and desk height (82.3%), followed by upper edge of backrest (47.1%), seat depth (25%), seat width (4.4%), and seat to desk clearance (1.47%).

Pain prevalence among pupils was high for most body regions. Knee pain was the most common musculoskeletal pain, reported by 31.1% of pupils. For other body regions pain prevalence was similar: upper back pain (26.4%), neck and lower back pain (23.5%), shoulder pain (22.1%), and ankle pain (21.0%). Wrist and elbow pain were less frequently reported (14.7 and 4.4%, respectively).

Logistic regression revealed no significant impact of school furniture on pain occurrence among pupils. However, the data obtained indicated upper back pain might be associated with unsuitable backrest.

Table 1 Participants characteristics

	Pupils
Age (mean ± SD)	12.1 ± 1.1
Body height (cm) (mean ± SD)	157.1 ± 0.1
Body mass (kg) (mean ± SD)	47.6 ± 11.6
Body mass index (kg/m^2) (mean ± SD)	19.1 ± 3.1

Table 2 Pupils' body dimensions

	Popliteal height	Thigh thickness	Elbow height sitting	Shoulder height sitting	Subscapular height sitting	Hip width	Buttock-popliteal length
	(M ± SD)	(M ± SD)	(M ± SD)	(M ± SD)	(M ± SD)	(M ± SD)	(M ± SD)
Pupils	42.3 ± 3.1	11.9 ± 1.8	19.8 ± 3.2	50.9 ± 4.2	39.0 ± 3.7	33.2 ± 4.0	45.3 ± 3.3

4 Discussion

The results of this study revealed that there is a considerable mismatch between school furniture and body dimensions of pupils. Particularly, seat and desk height were the most unsuitable for most of pupils included. Pain prevalence was also high with the most affected areas being knees, upper back, neck, lower back, and shoulders. On the contrary to our hypothesis the calculated mismatch was not associated with pain occurrence, although there was a trend associating upper back with unsuitable backrest.

Studies conducted so far reported ambiguous conclusions. Paradoxical results were reported by Brewer et al. [26] who found elbow-desk height mismatch to have a protective role against back discomfort and seat depth mismatch to protect from lower back discomfort. On the other hand, Ben Ayed et al. [23] reported too low desk to be associated with neck pain and too low seat backrest with lower back pain. Similarly, Murphy et al. [27] reported to low chair to be associated with neck, upper and lower back pain. However, in this study the mismatch was evaluated by questionnaire and was not objectively measured. Conflicting results from studies could be partly explained by different sample size characteristics in studies, where Brewer et al. [26] included 139 pupils from primary school, Ben Ayed et al. [23] included 1221 adolescents form primary and secondary school and a sample of 68 pupils from primary school took part in our study. Since it has been shown that musculoskeletal pain increases with age [4, 23] it is possible that influence of unsuitable school furniture is more pronounced among older students. Another difference among studies is also statistical analysis of the data. Although logistic regression has been used in majority of studies [23, 24, 26, 27], different approaches and adjustments in the statistical method have been used.

To elaborate on this issue further, dimensional incompatibility is only one of the several risk factors that affect musculoskeletal health of pupils in school. To diminish musculoskeletal conditions a comprehensive strategy is needed. In this context the policy-based approaches are desired to ensure the healthy environments for the pupils in schools.

Certain limitations should be considered when interpreting the results of the study. The study design was cross-sectional which cannot account for long-term changes, such as cumulative effects of student-furniture mismatch. The sample population was a convenience sample from one primary school. Also, the participants were not randomly selected and consequently, participants with musculoskeletal conditions were more likely to participate in the study.

5 Conclusion

This study provides an insight into the suitability of school furniture in Slovenian primary school in relation to pupils' body dimensions. A high mismatch has been recognised, particularly for seat and desk height. However, based on this study, the student-furniture mismatch does not significantly contribute to the pain development although the pain prevalence among student was high. Further research is needed to better understand the long-term effects of unsuitable school furniture on musculoskeletal conditions among pupils.

References

1. Woolf AD, Erwin J, March L (2012) The need to address the burden of musculoskeletal conditions. Best Pract Res Clin Rheumatol 26:183–224
2. Hertzberg A (1985) Prediction of cervical and low-back pain based on routine school health examinations a nine-to twelve-year follow-up study. Scand J Prim Health Care 3:247–253. https://doi.org/10.3109/02813438509013957
3. Domljan D, Vlaović Z, Grbac I (2010) Muskulockeletal deformities and back pain in school children. In: Proceedings of the 4th ergonomics conference: ergonomics 2010; Mijović B (ed) Stubičke Toplice, Zagreb: Croatian society of ergonomics, 2010, pp 227–236
4. Kristjansdottir G, Rhee H (2002) Risk factors of back pain frequency in schoolchildren: a search for explanations to a public health problem. Acta Paediatr 91:849–868154
5. King S, Chambers CT, Huguet A, MacNevin RC, McGrath PJ, Parker L, MacDonald AJ (2011) The epidemiology of chronic pain in children and adolescents revisited: a systematic review. Pain 152:2729–2738. https://doi.org/10.1016/j.pain.2011.07.016
6. Talbott NR, Bhattacharya A, Davis KG, Shukla R, Levin L (2009) School backpacks: it's more than just a weight problem. Work 34:481–494. https://doi.org/10.3233/wor-2009-0949
7. Lindstrom-Hazel D (2009) The backpack problem is evident but the solution is less obvious. Work 32:329–338. https://doi.org/10.3233/wor-2009-0831
8. Perrone M, Orr R, Hing W, Milne N, Pope R (2018) The impact of backpack loads on school children: a critical narrative review. Int J Environ Res Public Health 15(11):2529
9. Gunzburg R, Balagué F, Nordin M, Szpalski M, Duyck D, Bull D, Mélot C (1999) Low back pain in a population of school children. Eur Spine J 8:439–443. https://doi.org/10.1007/s005860050202
10. Reis J, Flegel M, Kennedy C (1996) An assessment of lower back pain in young adults: implications for college health education. J Am Coll Health Assoc 44:289–293. https://doi.org/10.1080/07448481.1996.9936857
11. Grimmer K, Williams M (2000) Gender-age environmental associates of adolescent low back pain. Appl Ergon 31:343–360. https://doi.org/10.1016/s0003-6870(00)00002-8
12. Steene-Johannessen J, Hansen BH, Dalene KE, Kolle E, Northstone K, Møller NC, Grøntved A, Wedderkopp N, Kriemler S, Page AS et al. (2020) Variations in accelerometry measured physical activity and sedentary time across Europe-harmonized analyses of 47,497 children and adolescents. Int J Behav Nutr Phys Act 17:38 https://doi.org/10.1186/s12966-020-00930-x
13. Castellucci HI, Arezes PM, Molenbroek JFM, de Bruin R, Viviani C (2017) The influence of school furniture on students' performance and physical responses: results of a systematic review. Ergonomics 60:93–110. https://doi.org/10.1080/00140139.2016.1170889

14. Alibegović A, Hadžiomerović AM, Pašalić A, Domljan D (2020) School furniture ergonomics in prevention of pupils' poor sitting posture. Drv Ind 71:88–99. https://doi.org/10.5552/drvind.2020.1920
15. Molenbroek JFM, Kroon-Ramaekers YMT, Snijders CJ (2003) Revision of the design of a standard for the dimensions of school furniture. Ergonomics 46:681–694. https://doi.org/10.1080/0014013031000085635
16. Agha SR (2010) School furniture match to students' anthropometry in the Gaza Strip. Ergonomics 53:344–354. https://doi.org/10.1080/00140130903398366
17. Assiri A, Mahfouz A, Awadalla N, Abouelyazid A, Shalaby M, Abogamal A, Alsabaani A, Riaz F (2019) Classroom furniture mismatch and back pain among adolescent school-children in Abha City, Southwestern Saudi Arabia. Int J Environ Res Public Health 16:1395. https://doi.org/10.3390/ijerph16081395
18. Baharampour S, Nazari J, Dianat I, Asgharijafarabadi M (2013) Student's body dimensions in relation to classroom furniture. Heal Promot Perspect 3:165–174. https://doi.org/10.5681/hpp.2013.020
19. Castellucci HI, Catalán M, Arezes PM, Molenbroek JFM (2016) Evaluation of the match between anthropometric measures and school furniture dimensions in Chile. Work 53:585–595. https://doi.org/10.3233/wor-152233
20. Cotton LM, O'Connell DG, Palmer PP, Rutland MD (2002) Mismatch of school desks and chairs by ethnicity and grade level in middle school. Work 18:269–280
21. Gouvali MK, Boudolos K (2006) Match between school furniture dimensions and children's anthropometry. Appl Ergo 37:765–773. https://doi.org/10.1016/j.apergo.2005.11.009
22. Domljan D, Grbac I, Hađina J (2008) Classroom furniture design—correlation of pupil and chair dimensions. Coll Antropol 32:257–265
23. Ayed HB, Yaich S, Trigui M, Hmida MB, Jemaa MB, Ammar A, Jedidi J (2019) Prevalence, risk factors and outcomes of neck, shoulders and low-back pain in secondary-school children. J Res Heal Sci 19(1):e00440
24. Gheysvandi E, Dianat I, Heidarimoghadam R, Tapak L, Karimi-Shahanjarini A, Rezapur-Shahkolai F (2019) Neck and shoulder pain among elementary school students: prevalence and its risk factors. BMC Public Health 19:1299. https://doi.org/10.1186/s12889-019-7706-0
25. Castellucci HI, Arezes PM, Molenbroek JFM (2014) Applying different equations to evaluate the level of mismatch between students and school furniture. Appl Ergon 45:1123–1132. https://doi.org/10.1016/j.apergo.2014.01.012
26. Brewer JM, Davis KG, Dunning KK, Succop PA (2009) Does ergonomic mismatch at school impact pain in school children? Work 34:455–464. https://doi.org/10.3233/wor-2009-0946
27. Murphy S, Buckle P, Stubbs D (2007) A cross-sectional study of self-reported back and neck pain among English schoolchildren and associated physical and psychological risk factors. Appl Ergon 38:797–804. https://doi.org/10.1016/j.apergo.2006.09.003

Hammer Design Using Ergonomic Knowledge-Based Guidelines

Nebojsa Rasovic, Adisa Vucina, and Diana Milčić

Abstract Manual hand tool design must meet all requirements making usable, safe and comfortable hand tool. An effective plan of design guidelines must define and implement all aspects of influence factors at the same time. Advances and innovativeness in tool design nominate complex geometry to achieve a better interaction between user and its tool. The goal of the paper is to create and evaluate an ultimate hand tool design preventing inefficiency and musculoskeletal disorders. A case study was conducted on the hammer.

Keywords Ergonomics · Hand tools · Product design and development

1 Introduction

Improper use or poor design of manual hand tools used in occupational work situations can cause discomfort, various injuries or musculoskeletal disorders affecting upper limbs and neck pain [1]. These disorders can also affect nerves, bones, joints and muscles in any part of the human body. To develop an efficient hand tool, the designer has to take into account the primary tool function, working environment and processing materials and the gender of users. In order to reduce discomfort and injury risk, it is crucial to use ergonomically designed hand tools. Ergonomics is an applied science concerned with designing and arranging elements of systems humans use so that the humans and other elements of the system interact most efficiently and safely. Guidelines for the ergonomic design of manual hand tools must include weight, shape, and dimensions to fit the user hand [2]. In order to make safe and comfortable manual hand tools, designers also have to consider the

---

N. Rasovic (✉) · A. Vucina
University of Mostar, Faculty of Mechanical Engineering, Computing and Electrical Engineering, Matice hrvatske bb, 88000 Mostar, Bosnia and Herzegovina
e-mail: nebojsa.rasovic@fsre.sum.ba

D. Milčić
University of Zagreb, Faculty of Graphic Arts, Getaldiceva 2, 10000 Zagreb, Croatia

D. Sumpor et al. (eds.), *Proceedings of the 8th International Ergonomics Conference*, Advances in Intelligent Systems and Computing 1313,
https://doi.org/10.1007/978-3-030-66937-9_20

size, shape and layout of the workstation and the way of the task performing. Therefore, it is crucial to address all these aspects to reduce the risk of work-related musculoskeletal disorders and accidents.

Having in mind the ergonomics of a hand tool, in addition to performing its primary function, the most critical part of the tool is its handle. Research into the anthropometry of the human hand can be used for selection of hand tools, but also provide a reasonable basis for guidelines for proper sizing of hand tools which will ensure comfortable operation [2–4]. Also, care must be taken of the material from which the hand tools are made. Unsuitable materials can contribute to injuries and accidents [5]. The texture of the tool handle is also a crucial design aspect regarding the adjustment of hand position [6].

In this study, authors are dealing with prevention of ergonomic discomforts and injuries which occurs using a traditional hammer. The use of simple hammers dates to more than 3 million years ago. The first hammers were made without handles. It is believed that humans achieved significant technical progress when they put the handle on the hammer and thus increased the strength and precision of impact. Stones attached to sticks with strips of leather were used as hammers with handles by about 30 thousand years ago. It took thousands of years to create such revolutionary progress in hammer design [5].

From history to present day, the hammer has been a simple manual hand tool whose primary function is to amplify muscles forces of impact by converting mechanical work into kinetic energy. The mechanical advantage provided by increasing moment arm ratios, greater kinetic energy, produces mechanical work not possible with a bare hand. It consists of a weighted head fixed to a long handle that is swung to deliver an impact to a small area of an object. The existing non-powered hammerhead is typically made of steel, and the handle is typically made of wood or plastic in order to absorb shocks and reduce user fatigue from repeated strikes. The hammer is one of the most commonly used hand tools, both in the household, agriculture and in various industries. Although hammer can protect the bare hand from mechanical injury, improper use or poor design of the hammer in occupational work situations may introduce stresses through the hand-holding or operating it. That cause musculoskeletal disorders and injuries such as shoulder pain, wrist pain, soreness in elbow, stiffness in finger and headache [7].

For comfortable use without pain and injures, hammer with ergonomic features is continuously redesigned, especially over the past decade. Grip cross-section shape, grip size, and handle length are important geometric parameters in manual hand tool design. It is essential to prevent or reduce wrist flexion, extension or deviation is to keep the wrist straight or in a neutral position. To address this challenge, it seeks harmonization of the ergonomics, anthropometrics and biomechanics of the hammer with the anatomy of the human hand.

In this paper, ergonomic guidelines for hammer design will be defined, and a new hammer will be designed according to these guidelines. It is planned to make 3D printed plastic prototype of a new hammer and evaluate the design in the term of ergonomics.

2 Ergonomic Design of Hammer Handle

Handle weight, form and dimensions have to be designed to avoid wrist, elbow and shoulder musculoskeletal disorders. Good grip design provides the user with better control and fewer accidents. Many studies have been focused on defining optimal handle design based on the multi-criteria evaluation [8]. These criteria and parameters are very variable, and they depend mostly by user anthropometry, tool function and working conditions.

With aims at collecting, organizing and applying basic ergonomic guidelines for hammer design, ergonomic, anthropometric and biomechanical research data are integrated into the study (Table 1). Tool design (weight, shape, fit to the user) along with tasks to be performed are critical factors in making hand tool, that should be using safe and risk-free. An efficient and active prevention strategy must address all mentioned aspects simultaneously.

2.1 Hammer Form and Dimensions

In this study, grip size (GS), grip diameter (GD) and handle length (L) are determined to design hammer for 50 percentiles of European population [10]. A power grip is a grip formed with the fingers and the palm to use a hammer. Grip size is determined by the circumference around the edge of the handle. This circumference is measured in the middle of the handle. Equations (1) and (2) are used to determine grip size (GS) and grip diameter (GD). The length of the hand (HL) is measured between the wrist crease and the tip of the most extended finger on the hand, usually, thumb finger. Handbreadth (HB) is the length of the palm, measured

Table 1 Ergonomic guidelines for hammer design [9]

Description	Guideline	Reason
Direction of force is perpendicular to forearm and wrist (typically vertical)	Straight handle	Minimal wrist deviation
Handle shape	Slightly contoured	Easy-grip
Handle length	100–125 mm	Keep contact out of the palm
Handle cross-section	Round or oval	Easy fit and grip
Handle diameter (power grip)	30–50 mm	Greater force and stability
Material and texture of handles	Non-slip, non-conductive materials	For comfort and reduces the effort required to use tool

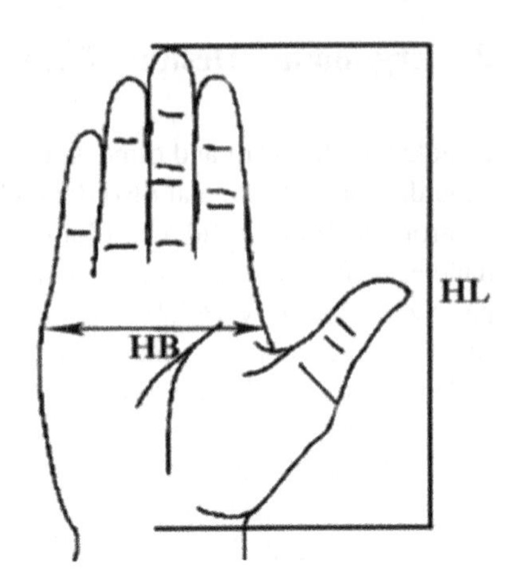

Fig. 1 Hand anthropometry [7]

perpendicular to hand length (Fig. 1). Optimal handle length (L) according to ergonomic recommendations (Table 1) should be in a range of 100–125 mm. A handle that is too short may cause unnecessary compression in the middle of the palm. Most workers have the breadth of the palm less than 100 mm [1] and considering gloves width, we have determined that hammer handle should be not less than 100 mm to reduce the harmful effects of any compression exerted.

$$GD = 0.2HL \quad (1)$$

$$GS = GD\pi \quad (2)$$

Since recent research on biomechanics recommend an oval cross-section of hand tool handle with a length-to-width ratio of 1:1.25, an oval cross-section with a grip size of 115 mm has been designed. Optimal handle diameter according to ergonomic recommendations (Table 1) should be in a range of 30–50 mm. Figure 2 shows handle cross-sections of the ergonomic hammer.

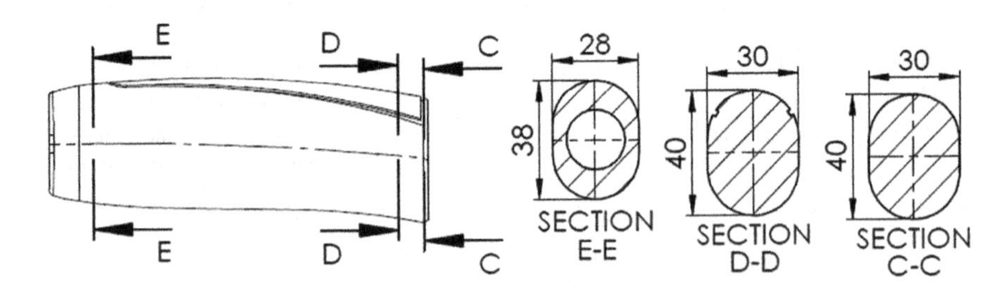

Fig. 2 Cross-sections of ergonomic hammer handle design

2.2 Contact Pressure

Contact pressure occurs on the inner side of the palm caused by mechanical stress in multi-axis direction. We have several stress conditions as twisting, bending and compression. To avoid contact pressure or to minimize it, the soft rubber filler is designed into the handle body. Local pressure occurs on the tissue of the palm, causing nerves damage. This pressure is caused by the tight fit of the user's palm with a rigid handle surface under great forces exerting. Guidelines for ergonomic hammer design are defined as follows:

- Choose hammer with a handle that has added friction such as compressible rubber or closed-cell foam, with slightly etched surfaces.
- Use contoured handle which the pressure spread over a larger area.
- Use the handle of adequate length and thickness that span the entire hand.

The soft rubber filler has been designed to prevent this kind of pressure or to minimize it. The curved longitudinal geometry of the handle, including appropriate length and cross-sectional measurements, have been used to spread pressure over a wider area. The curved contour also ensures a tighter grip and prevents the hand from slipping off the handle when performing a stroke.

2.3 Vibrations

The vibrations can cause discomfort to the human nervous system [8]. Thus the distance must be more generous as well as possible between vibration-generating source and the human hand. Various tool covers and anti-vibration gloves are recommended as safeguard measures in the design process. Regarding vibration elimination following guidelines for hammer design are defined:

- Avoid or limit vibration. Select the hammer that minimizes vibration.
- Use isolation mounts such as springs and silent rubber blocks between individual parts.
- Use damping materials on a handle's surface

To avoid vibrations or to minimize it, the rubber cover has been designed at the joining place between the handle and bar.

2.4 Reducing Excessive Forces

The excessive forces are conducted to the fingers generating the stress on tendons, muscles, wrist and up to the carpal tunnel. This kind of forces stretch and exhaust tendons. If it comes to the combination of repetitive and prolonged motions, it may lead to injuries and carpal tunnel syndrome. Regarding excessive forces reducing the following guidelines for hammer design are defined:

- Reduce excessive gripping force or pressure.
- Avoid high contact forces and static loading.
- Avoid extreme and awkward wrist and arm positions.
- Avoid sharp edges on the handle.
- Use the appropriate grip size to generate the greatest grasping force, design grip to a size that permits the thumb and forefinger to overlap slightly.
- Improve the overall hammer design (improve the balance between dimensions and mass).

Following ergonomic guidelines and principles, a new hammer handle has been designed. New dimensions, which are resulting from the collected-integrated data and the modelling process, are shown in Fig. 3. The virtual model is graphically rendered to describe design intention—the graphic highlights major ergonomic features to confirm the adopted recommendations.

3 Design Evaluation

Evaluation of the ergonomic hammer and existing hammers was conducted among 20 industry male-workers who frequently use a hammer. They had the opportunity to try the prototype of the ergonomically designed hammer, after which a survey was conducted. Table 2 presents the anthropometric measures of surveyed users.

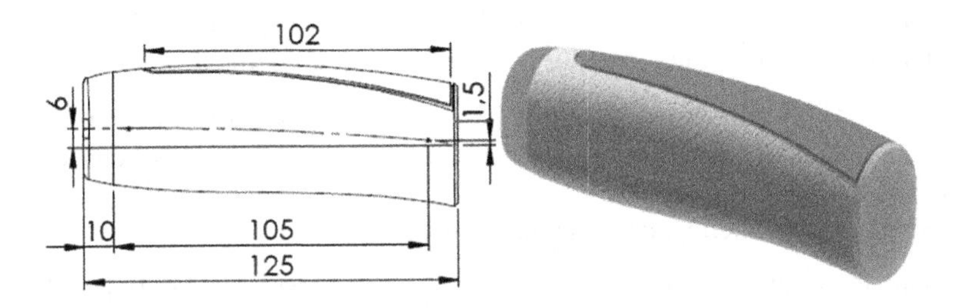

Fig. 3 Ergonomic hammer handle design and virtual model

Table 2 Anthropometric measures of surveyed hammer users

	Age	Hand breadth (mm)	Hand length (mm)
Average	40.5	90.3	192.1
Max	64	102	203
Min	23	77	178

Table 3 Criteria for experimental grip size testing

Evaluation	Description
Grip size is too small	Two middle fingers are digging into the heel portion of the user's palm
Grip size is correct	Two middle fingers are slightly gapped to touching the heel portion of the user's palm, thumb and forefinger slightly overlap
Grip size is too large	Two middle fingers have too much gap to the heel portion of the user's palm

3.1 Grip Size Evaluation

Too small grip size would require more hand strength to prevent the hammer rod from twisting in hand, which could result in discomfort and musculoskeletal disorders. On the contrary, if there is a large gap between middle fingers and the palm of the hitting hand, the grip size is too big. Ultimately, it would be best to achieve that the ring finger and middle finger are close to the user's palm (Table 3).

Testing shows that most of the surveyed workers for this study are not satisfied with existing hammers and want a new hammer because they are not comfortable with existing ones. Grip size of existing hammers has been measured, and it was in the range from 80 to 100 mm, 88 mm on average.

Visual testing and measuring the gap between middle fingers and the heel of the user's palm for 20 workers shows that grip size of the new ergonomic hammer is correct and for existing hammers is too small (Figs. 4 and 5).

3.2 Comfort Evaluation

Since only the 3D printed plastic prototype of the ergonomically designed hammer is made for this evaluation, it was not possible to examine and compare all design

Fig. 4 Grip size testing of the new ergonomic hammer handle

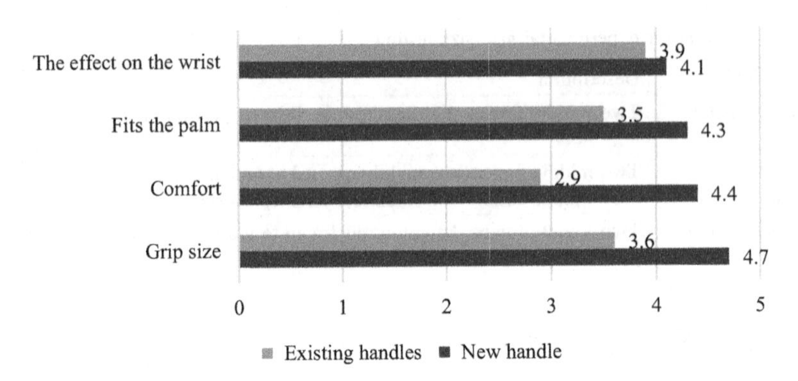

Fig. 5 Evaluation results

characteristics related to handling flexibility and user comfort mentioned in guidelines above. Unfortunately, at this stage, it was not possible to test some vital grip characteristics such as surface finish, handle quality and effects of rubber filler. Comfort characteristics such as 'Overall comfort at first look', 'Fits the palm' and 'Effect of hammer use on the wrist', have been used as criteria for comfort analysis. Results of the evaluation are presented in Fig. 5.

The horizontal chart axis shows the average mark of the conducted evaluation among 20 male-workers by the set comfort criteria. Each worker could answer each comfort criteria on the scale from 0 to 5. Four existing hammer handles along with the new hammer handle have been evaluated on workers belonging to the 50 percentile group.

4 Conclusion

Implementation of ergonomic knowledge-based guidelines and principles in new hammer design significantly helps to improve existing hammer in terms of comfort and safe manual operation. Ergonomic guidelines and principles defined in this paper present a valuable knowledge which can be transformed into a functional ergo-design going beyond conventional design. First testing and evaluation proved ergonomic design of the new hammer. Grip size is correct, overall comfort and better effects on the wrist are confirmed. Further testing of the new ergonomic hammer has to be done on a pre-produced ergonomic hammer to ensure evaluation of other essential ergonomic aspects including surface finish, handle quality and weight balance. It would be good to focus future research on the need to design a hammer family for more percentile groups with the aim of better adapting to different genders and hand sizes.

References

1. Putz-Anderson V (1992) Cumulative trauma disorders: a manual for musculoskeletal diseases of the upper limbs. Taylor & Francis, London
2. Wang C, Cai D (2017) Hand tool handle design based on hand measurements. In: MATEC web of conferences, vol 119, p 1044
3. Chandra A, Chandna P, Deswal S (2011) Analysis of hand anthropometric dimensions of male industrial workers of Haryana State. Int J Eng 5(3):242–256
4. Mohammad YAA (2005) Anthropometric characteristics of the hand based on laterality and sex among Jordanian. Int J Ind Ergon 35(8):747–754
5. Parmeggiani L (ed) (1983) Encyclopaedia of occupational health and safety, pp 24–26. International Labour Office, Geneva
6. Lewis WG, Narayan CV (1993) Design and sizing of ergonomic handles for hand tools. Appl Ergon 24(5):351–356
7. Haque M (2018) Ergonomic design of hammer handle to reduce musculoskeletal disorders of carpenters. Int J Eng Technol 4(2):78–83
8. Kaljun J, Dolšak B (2012) Ergonomic design knowledge built in the intelligent decision support system. Int J Ind Ergon 42(1):162–171
9. Hand Tool Ergonomics–Tool Design (2015) Canadian Centre for Occupational Health & Safety. https://www.ccohs.ca/oshanswers/ergonomics/handtools/tooldesign.html
10. Jurgens HW, Matzdorff I, Windberg J (1998) International anthropometric data for work-place and machinary design. Dortmund, Germany

Functional Networks for Modeling and Optimization Human-Machine Systems

Evgeniy Lavrov, P. Paderno, O. Siryk, V. Kyzenko, S. Kosianchuk, N. Bondarenko, and E. Burkov

Abstract The article examines the issues of reliability of human-machine systems and modeling of human-machine interaction. The classification of models for describing human activity and its interaction with a machine is carried out. The advantages of functional networks for ergonomics are shown. The description of the mathematical apparatus of functional networks is given. The characteristics of the models for describing, evaluating and optimizing the algorithms of the human operator's activity when they work with automation are given. Examples of models for control systems for complex objects of different types are given. A computer system for modeling and optimizing the activities of operators of control systems is described. The tasks are set for improving the theory and practice of functional networks, associated with the development of situational control models and intelligent analysis of data on the reliability of the human operator. The results will be useful for finding ergonomic reserves for increasing the reliability of automated systems.

Keywords Reliability · Simulation · Ergonomics · Man-machine system · Activity algorithm · Optimization

E. Lavrov (✉)
Sumy State University, Sumy, Ukraine
e-mail: prof_lavrov@mail.ru

P. Paderno · E. Burkov
St. Petersburg Electrotechnical University "LETI", St. Petersburg, Russia

O. Siryk
Taras Shevchenko National University, Kyiv, Ukraine

V. Kyzenko · S. Kosianchuk · N. Bondarenko
Institute of Pedagogy of the National Academy of Educational Sciences of Ukraine, Kyiv, Ukraine

D. Sumpor et al. (eds.), *Proceedings of the 8th International Ergonomics Conference*, Advances in Intelligent Systems and Computing 1313,
https://doi.org/10.1007/978-3-030-66937-9_21

1 Introduction

Despite the fourth industrial revolution and progress in the development of artificial intelligence, the expected complete exclusion of man from the field of computerized production did not occur [1, 2]. Moreover, the importance of his role and the intensity of his activities in this area in connection with the growth of responsibility and the cost of mistakes are constantly increasing [3, 4].

2 Problem Statement

The catastrophic consequences of ignoring the human factor for critical systems [5, 6] actualize the task of designing ergonomic measures [7, 8]. The traditional problems of ergonomics are [9–11]: anthropometry, biomechanics, psychophysiology, perception of information, memory, thinking, safety, prevention of musculoskeletal disorders and risk of various diseases, etc. However, a number of scientists [10, 12–14] notes that despite the importance of these and other local tasks, in the conditions of a new stage in the development of ergonomics—«ergonomics of the information space»—complex tasks of organizational ergonomics come to the fore, that is, the tasks of modeling and optimizing the activities of operators taking into account all significant factors. In this regard, we define the goal of this work: to describe the methodology of modeling by functional networks of operators' activities (little presented in the English-language scientific literature), taking into account the structures of human-machine interaction and the entire range of influencing factors.

3 Results

3.1 History and Methodology of Modeling Human-Machine Systems by Functional Networks

The founder (1963) of the functional-structural theory [14–16] of ergo-technical systems was Anatoly Ilyich Gubinsky (1931–1990), a military sailor, Doctor of Technical Sciences, Professor, Captain 1st Rank, the first president of the Soviet Ergonomic Association. The impetus for his research in the field of ergonomics was the experience gained during his service in the navy: in 1959 A.I. Gubinsky took part in the first voyage of a nuclear submarine cruiser to the North Pole. By 1981, under the leadership of professor A.I. Gubinsky, the scientific direction «Efficiency, quality and reliability of human-technology systems» was finally formed [14–16].

Ergonomic associates (doctors of sciences) Vladimir Evgrafov, Yuri Grechko, Valentin Kobzev, Evgeny Tsoi, Mikhail Grif, Evgeny Pavlov (Russia); Akiva

Asherov, Pavel Chabanenko, Alexander Burov, Georgy Kozhevnikov, Sergey Shtovba (Ukraine); Alexander Rothstein (Israel), Marjan Wlodarczyk (Poland) and many others not only participated in the development of principles and models for the evaluation and optimization of man-machine systems, but also tested them on systems of various purposes and in various fields [13, 16, 17]:

- marine fleet;
- aviation and astronautics;
- control systems for technological processes of various types;
- enterprise management systems;
- training systems.

To model and evaluate such logically complex systems, it is in principle impossible or difficult to use existing network and other models [18, 19], for example:

- logical systems (formal grammars, Petri networks, etc.);
- algebraic systems (Markov and semi-Markov processes);
- language-algebraic systems (antecedent networks, PERT, GERT, etc.).

This is due to the fact that these models have common drawbacks: the impossibility of describing the complex logic of human interaction with technology and (or) the impossibility of obtaining an assessment of the reliability and timeliness of the performance of activity processes. Therefore, a fundamentally new type of model was developed—a functional network (FN), which describes and allows to evaluate the process of human-machine interaction. FN is a work graph, the vertices of which correspond to operations, and the arcs correspond to relations between operations, and (or) an event graph (events describe, for example, the processes of occurrence and elimination of errors or failures). To describe the process of interaction, typical functional units (TFU) are used: functionaries that correspond to real operations and actions, and composers necessary to establish logical and functional links between functionaries.

The introduction into the alphabet of the FN apparatus of operations of control of functioning and operability allows, in contrast to other network methods, to simulate the processes of loss of stability of functioning due to errors and failures. FN can describe cyclic processes (loops), both with a limitation on the number of repetitions, and without this limitation. The logic for executing parallel operations is wider than in GERT, and allows to implement the functions of logic algebra «AND», «OR», «XOR» both at the input and at the output of the parallel structure. To assess the reliability (accuracy, timeliness) and resource consumption, a set of analytical expressions was obtained for the most common structures, the so-called typical functional structures (TFS). If in 1991 the base of models contained models for 18 TFS, then in 2020 it contains models for 55 TFS.

An assessment of the quality of the execution of the entire algorithm of human-machine interaction is obtained by sequential iterative combining of TFU into TFS (by «folding» of TFU).

3.2 Solution of Ergonomics Problems by Methods of Functional Networks. Automating Modeling in Ergonomics

After the death of A.I. Gubinsky in 1990, the collapse of the USSR and the general crisis in science, work within the framework of the scientific school was suspended. In 1997, we resumed research within the framework of Russian-Ukrainian cooperation in connection with the urgent need to develop decision support systems to find ergonomic reserves for the effectiveness of new generation automation systems. The effectiveness of assessment and optimization models based on FN for solving typical problems of ergonomics has been proven [16, 17, 20]:

- Er_1—distribution of functions between man and technology (choice of the degree of automation);
- Er_2—determination of the number of personnel and the distribution of functions between operators;
- Er_3—design of information models (interfaces);
- Er_4—design of operator activity algorithms;
- Er_5—design of working conditions and certification of workplaces.

A need arose to solve a number of fundamentally new problems at a new level, due to the fact that in order to make decisions on-line, FN must be analyzed quickly and automatically, excluding human participation. To do this, a new computer-optimized FN description language and its automatic parsing method were developed [20, 21]. This and a number of other new models form the basis of the simulating qualimetric complex of human-machine interaction (Table 1).

Table 1 The main subsystems of the simulating qualimetric complex

Subsystem	Model types	Solvable tasks	Source
1. Storage (formation, taking into account influencing factors) of the initial data on the reliability and duration of operations	Databank, predictive model (neural network, fuzzy logic, etc.)	Er_1–Er_4	[21]
2. FN assessment, variant analysis	TFU, TFS, FN description language, FN analysis mode	Er_1–Er_4	[21]
3.FN optimization	Single-criteria and multi-criteria tasks (task bank contains 56 models)	Er_1–Er_4	[22]
4.Assessment of physical and psychological factors of the environment, assessment of the workplace	Normative, expert, fuzzy models; integral scores	Er_5	[23]
5. Ergonomic expertise	Bank of models for the formation and processing of expert assessments	Er_1–Er_5	[24]

An example of the results of calculating reliability (for a flexible production system), performed using the software package, is shown in Fig. 1.

On the left, Fig. 1 shows the algorithm of the operator's activity using the symbols accepted in the scientific school of A.I. Gubinsky [14–17]: rectangle—working operation (mandatory for a technological process), circle—an operation of checking the correctness of the technological process (introduced to increase error-free). On the right, Fig. 1 shows how a specially designed computer program identifies typical functional structures in the activity algorithm, substitutes human reliability characteristics into the corresponding formulas, and replaces the found structures with single operations with equivalent characteristics (Pe_i designating an equivalent operation that replaced a certain typical functional structure in the i-th step). The presented process of folding typical structures consists of 8 steps. In the last step, the desired error-free probability values and runtime statistics are obtained. All necessary calculations are carried out completely automatically, which allows an ergonomist or manager not to waste time on them. Assessments can be obtained for different initial data (taking into account differences in equipment, working conditions, operator skills, etc.) and different activity structures, while providing a graphical display of existing dependencies and the possibility of choosing a rational option.

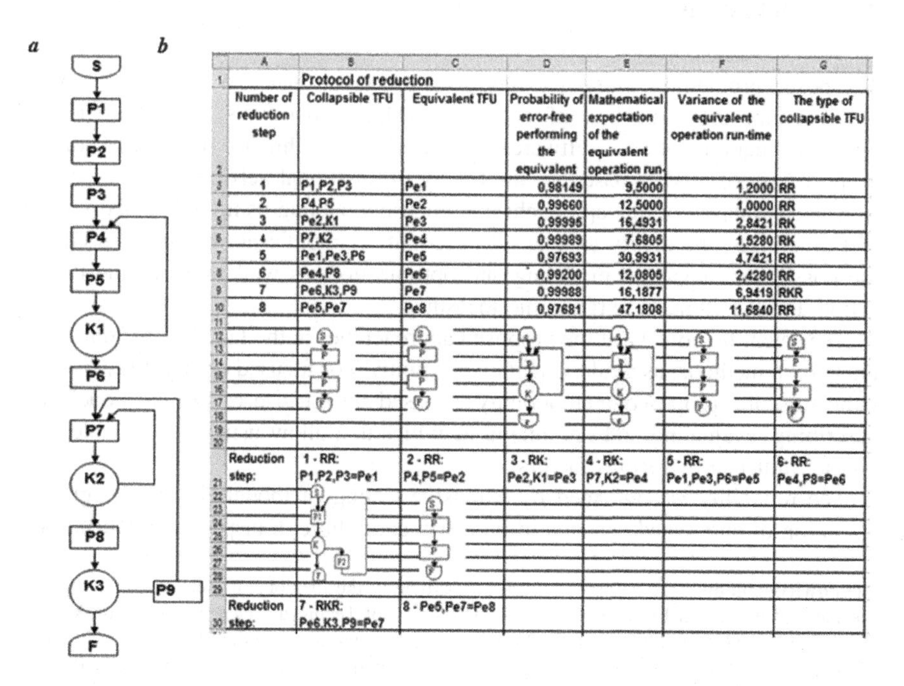

Protocol of reduction

Number of reduction step	Collapsible TFU	Equivalent TFU	Probability of error-free performing the equivalent	Mathematical expectation of the equivalent operation run-	Variance of the equivalent operation run-time	The type of collapsible TFU
1	P1,P2,P3	Pe1	0,98149	9,5000	1,2000	RR
2	P4,P5	Pe2	0,99660	12,5000	1,0000	RR
3	Pe2,K1	Pe3	0,99995	16,4931	2,8421	RK
4	P7,K2	Pe4	0,99989	7,6805	1,5280	RK
5	Pe1,Pe3,P6	Pe5	0,97693	30,9931	4,7421	RR
6	Pe4,P8	Pe6	0,99200	12,0805	2,4280	RR
7	Pe6,K3,P9	Pe7	0,99988	16,1877	6,9419	RKR
8	Pe5,Pe7	Pe8	0,97681	47,1808	11,6840	RR

Fig. 1 Operator's activity in managing an automated warehouse module of a flexible production system: **a** model, **b** process and results of automatic calculation

3.3 Tasks of Development of Theory and Practice of Modeling of Activity by Functional Networks

Prospects for the further development of the FN theory lie in the gradual removal of a number of assumptions and limitations, in particular, in the creation and development of:

- managed FN (for models of «flexible» activity and situational management);
- FN with queues (for polyergatic systems with flows of requests);
- FN with models of identification and control of the state of operators;
- FN with time-varying initial data (taking into account tempo tension, emotional state, working conditions, etc.);
- fuzzy FS (fuzzy structure of activity, fuzzy models of the formation of initial data, fuzzy optimization);
- progressive methods of entering information about the FN (voice input, input using a smartphone, etc.).

4 Conclusions

The posed problem of integrated modeling of human-machine interaction (taking into account the structures of activity; working conditions; characteristics of operators, equipment and software) can be solved within the framework of the proposed set of models based on the theory of functional networks.

The advantages of this approach: qualimetric modeling using indicators of error and timeliness, the ability to take into account ergonomic norms and restrictions, the possibility of variant modeling and optimization, as well as economic justification and the choice of ergonomic measures.

The scientific novelty of the proposed approach lies in the fact that, in contrast to the known methods, as a rule, qualitative (intuitive) or focused on the study of local indicators, the proposed methodology is based on the use of formal models for describing, evaluating and optimizing activities that allow using all available data on the influence of various factors.

The reliability is confirmed by extensive testing of models in the design and operation of automated systems in industry, agriculture, transport and education.

Acknowledgements The authors dedicate this article to the memory of their teacher – the first president of the Soviet Ergonomic Association, Doctor of Technical Sciences, Professor Anatoly Ilyich Gubinsky (1931–1990, St. Petersburg, Russia), who was the founder of the scientific school «Efficiency, quality and reliability of man-machine systems» and who first formulated the ideas that formed the basis of our study.

References

1. Kagermann H, Albach H. et al. (2015) Change through digitization-value creation in the age of Industry 4.0. In Management of permanent change. Springer, New York, pp 23-45
2. Sedova N, Sedov V, Bazhenov R (2018) Neural networks and fuzzy sets theory for computer modeling of ship collision avoidance in heavy traffic zone. In: 2018 International Russian automation conference (Rusautocon), sochi, pp 1–5 https://doi.org/10.1109/rusautocon.2018.8501752
3. Bloomfield R, Lala J (2013) Safety-critical systems. The next generation. In IEEE Secur & Priv 11(4): 11–13
4. Mary GMJ, Munipriya P (2011) Health monitoring of IT industry people. In: 3rd International Conference on electronics computer technology, Kanyakumari, pp 144–148. https://doi.org/10.1109/icectech.2011.5941875
5. Zhong RY, Xu X, Klotz E, et al. (2017) Intelligent manufacturing in the context of industry 4.0: a review. Engineering **3**(5):616–630
6. Cao Q, Qian Z, Lin Y, Zhang WJ (2015) Extending axiomatic design theory to human-machine cooperative products. In: 2015 IEEE 10th conference on industrial electronics and applications (ICIEA), Auckland, pp 329–334
7. Bentley TA, Teo STT, McLeod L, Tana F, Bosua R, Gloet M (2016) The role of organisational support in teleworker wellbeing: a socio-technical systems approach. Appl Ergon 52:207–215.https://doi.org/10.1016/j.apergo.2015.07.019
8. Xu MP, Wang J, Yang M, Wang W, Bai Y, Song, Y (2017) Analysis of operator support method based on intelligent dynamic interlock in lead-cooled fast reactor simulator. Ann Nucl Energy 99:279–282
9. Zhirabok AN, Kalinina NA, Shumskii AE (2018) Technique of monitoring a human operator's behavior in man-machine systems. J Comput Syst Sci Int 57(3):443–452
10. Cacciabue PC (2014) Human error risk management for engineering systems: a methodology for design, safety assessment, accident investigation and training. Reliab Engineering &Syst Saf 83(2): 229–269
11. Dul J et al (2012) A Strategy for human factors/ergonomics: developing the discipline and profession. Ergonomics 55(4):377–395
12. Havlikovaa M, Jirglb M, Bradac Z (2015) Human reliability in man-machine systems. Procedia Engineering 100:1207–1214
13. Grif MG, Ganelina NG, Kochetov SA (2018) Automation of human-machine systems design based on functional-structural theory. In: XIV International scientific-technical conference on actual problems of electronics instrument engineering (APEIE), Novosibirsk, pp 396–399. https://doi.org/10.1109/apeie.2018.8545934 (2018)
14. Adamenko AN, Asherov AT, Berdnikov IL, et al. (1993) Information controlling human-machine systems: research, design, testing. Reference book: Gubinsky AI, Evgrafov VG (eds) Moscow, Russia: Mashinostroenie (1993) (in Russian)
15. Popovich PR, Gubinskiy AI, Kolesnikov GM (1985) Ergonomic support of astronauts' activities. Mechanical Engineering, Moscow, Russia (In Russian)
16. Chabanenko PP (2012) Research of the safety and efficiency of the functioning of systems "human—technics" by ergonomic networks. Sevastopol, Academy of naval forces named after PS Nahimov (2012) (in Russian)
17. Grif MG, Sundui O, Tsoy EB (2014) Methods of designing and modeling of man–machine systems. Proc Int Summer Work Comput Sci 38–40
18. Drakaki M, Tzionas P (2017) Manufacturing scheduling using colored petri nets and reinforcement learning. Appl Sci 7(2):136–143. https://doi.org/10.3390/app7020136
19. Farah K, Chabir K, Abdelkrim MN (2019) Colored Petri nets for modeling of networked control systems. In: 19th international conference on sciences and techniques of automatic control and computer engineering (STA), Sousse, Tunisia, pp 226–230. https://doi.org/10.1109/sta.2019.8717215

20. Lavrov E, Pasko N, Siryk O, Burov O, Morkun N (2020) Mathematical models for reducing functional networks to ensure the reliability and cybersecurity of ergatic control systems. In: 2020 IEEE 15th International conference on advanced trends in radioelectronics, telecommunications and computer engineering (TCSET), Lviv-Slavske, Ukraine, pp 179–184. https://doi.org/10.1109/tcset49122.2020.235418
21. Lavrov E, Paderno PI, Burkov EA, Siryk OE, Pasko NB (2020) Information technology for modeling human-machine control systems and approach to integration of mathematical models for its improvement. In 2020 XXIII international conference on soft computing and measurements (SCM), St. Petersburg, Russia, pp 117–120. https://doi.org/10.1109/scm50615.2020.9198791
22. Lavrov E, Pasko N, Kozhevnykov G, Gonchar V, Mukoseev V (2018) Improvement for ergonomic quality of man-machine interaction in automated systems based on the optimization model. In: International scientific and practical conference «problems of infocommunications. science and technology, PICS & T, pp 735–741. https://doi.org/10.1109/infocommst.2018.8632074
23. Lavrov E, Pasko N, Siryk, O (2020) Information technology for assessing the operators working environment as an element of the ensuring automated systems ergonomics and reliability. In: IEEE 15th international conference on advanced trends in radioelectronics, telecommunications and computer engineering (TCSET), Lviv-Slavske, Ukraine, pp 570–575. https://doi.org/10.1109/tcset49122.2020.235497
24. Lavrov E, Paderno P, Burkov E, Volosiuk A, Lung VD (2020) Expert assessment systems to support decision-making for sustainable development of complex technological and socio-economic facilities. E3S Web of Conferences, 166, 11002. https://doi.org/10.1051/e3sconf/202016611002

Burnout and Workload Among Health Care Workers in Bulgaria

Rumyana Stoyanova and Stanislava Harizanova

Abstract This study aimed to estimate the relationship between workload and burnout syndrome among healthcare professionals working in training or non-training healthcare organizations in Bulgaria. The study was conducted using the web-based Bulgarian Version of the Boyko's Burnout Inventory, which contains 84 statements grouped in 12 symptoms and 3 phases, and questions related to the weekly workload. The data were exported directly to SPSS 17.0 statistical software and analyzed with descriptive statistics and Spearman's correlation coefficient r_S. The level of significance of 5% probability ($p < 0.05$) was adopted. Based on the responses to the Boyko's Inventory 29.0% of the respondents had a high level of emotional exhaustion, 51.6% had a high level of resistance, and 31.50% had a high level of strain. The high level of burnout was found among 27.4%. The average weekly working hours for all respondents were 43.78 ± 16.51 h. For specialists who have been diagnosed with burnout, the average weekly working time was 48.47 ± 13.72 h. For those who did not receive a positive diagnosis, the average weekly duty time was 37.72 ± 15.22 h. There were average correlations between the level of burnout and the weekly working hours ($r_s = 0.286$; $p = 0.001$).

Keywords Workload · Burnout · Healthcare professionals

1 Introduction

In healthcare, workload is associated with excessive work hours due to shortages in the quantity of workers, shift work, and overwork. These working conditions cause worker burnout, which may lead to safety problems in patient care [1]. From ergonomics standpoint, the most important factor in incidents and accidents is imbalance between workload and human capability and limitations.

R. Stoyanova (✉) · S. Harizanova
Medical University of Plovdiv, 15a Vasil Aprilov blvd., Plovdiv 4002, Bulgaria
e-mail: rumi_stoqnova@abv.bg

D. Sumpor et al. (eds.), *Proceedings of the 8th International Ergonomics Conference*, Advances in Intelligent Systems and Computing 1313,
https://doi.org/10.1007/978-3-030-66937-9_22

Khan et al. defined workload as the amount of work supposed to be done in a specific time [2]. There is a strong relationship between these two components—the amount of work and working time, because if the amount of work is significant in terms of volume, then the time required for its implementation will be more. Given that in healthcare it is difficult to design and measure all activities performed within the work process as well as the presence of intense mental activity and regular staff shortages, therefore workload is most easily measured by working hours a week.

According to the existing literature, workload is one of the main predictors of burnout. Some researchers prove that workload has a strong relationship with burnout [3–5]. Other studies found that workloads have significant relationship with emotional exhaustion and depersonalization—two of three dimensions of burnout. López-Núñez et al. have proved that personal accomplishment has a less weight on burnout [6].

The burnout arises when job demands exceed the capacity of employees [7, 8]. Generally, it has been noted that those professions that demand more interactions with people, generate burnout.

The aim of the study is to estimate the relationship between workload and burnout syndrome among healthcare professionals working in training or non-training healthcare organizations in Bulgaria, using the Boyko's Burnout Inventory.

Until now, most studies have used Maslach's questionnaire to assess burnout. The originality of our research is the application of another burnout assessment method with proven very good psychometric characteristics, which will confirm or refute the existing findings [9].

2 Methods

2.1 *Study Design and Context*

A cross-sectional survey was conducted using an Internet-based software platform for assessment of burnout. It is free-to-access and available on website: https://mu-burnout.com/. The present study was part of the Project of the Medical University HO-13/2018.

2.2 *Participants*

The participants included a diverse group of professionals, either physicians or other health professionals—nurses, midwives, clinical and radiology technicians and rehabilitators, working in training or non-training healthcare organizations in Bulgaria, regardless of the length of experience at the institution. All of them were eligible for participation in this study.

2.3 Data Collection and Questionnaire

A self-administered questionnaire was used, accompanied by a letter in which the goal of the study was briefly introduced.

The first section of the questionnaire covers the demographic characteristics such as gender, age, marital status and questions related to workload.

The second section identifies the presence and level of burnout syndrome among participants using V. Boyko's method for diagnostics of the severity of symptoms and the phases of formation and completion of the "occupational burnout" process [10]. Boyko describes the dynamics of "occupational burnout" differentiating three stages each of which is manifested in the form of four symptoms [11]:

1. STRAIN phase: the presence of strain is a precursor for the starting and the development of the mechanism of formation of the burnout syndrome. This phase is characterized by the following symptoms: experiencing traumatizing psychological situations, dissatisfaction with oneself, feeling of being "enclosed in a cage", and anxiety and depression.
2. RESISTANCE phase: the introduction of this phase as a separate one is conditional. Where one realizes the presence of strains, they strive to avoid the effect of the emotional factor by reducing their emotional reactions, which leads to inadequate selective emotional reactions; emotional and moral disorientation; enlarged area of saving emotions, and reduced fulfillment of the professional duties.
3. EXHAUSTION phase: it is characterized by general loss of energy and failure of the nervous system. Emotional deficit, emotional avoidance (isolation), isolation of the personality (depersonalization), and psychosomatic and psycho-vegetative disorders are observed during this phase.

The clarity and homogeneity of interpretation of the scales is a significant advantage of Boyko's method. Thus, the score obtained from this test can be easily compared to the results obtained using other psychological diagnostic techniques. All these benefits make it the appropriate tool for assessment of burnout in our study.

2.4 Data Analysis Methods

Data were collected in the period from July 01 to July 31, 2019. The data were exported directly to SPSS 17.0 statistical software and analysed with descriptive statistics, non-parametric Mann-Whitney U and Kruskal–Wallis H tests and Spearman correlation coefficient (r_s). The level of significance of 5% probability ($p < 0.05$) was adopted.

3 Results

A total of 124 valid questionnaires were collected by the respondents. Descriptive statistics for the sample are presented in Table 1.

Based on the responses to the Boyko's Inventory—29.0% ($n = 36$) of the respondents had a high level of emotional exhaustion, 51.6% ($n = 64$) had a high level of resistance, and 31.50% ($n = 39$) had a high level of strain. The high level of burnout was found among 27.4% ($n = 34$).

Non-parametric tests found that there were no significant gender and marital status differences in burnout. On Table 2 is presented the correlation between age, workload and burnout.

The correlation between "Average working hours a week" and "Work more than 5 days a week", and "Work more than 8 h a day" was strong. These results confirmed validity of instrument. The people who worked more than 8 h a day, more than 5 days a week had significantly more working hours a week.

Table 1 Demographic and work-related characteristics of the sample

Variable	Description	Mean ± SD	
		n	(%)
Gender	Male	35	28.2
	Female	89	71.8
Age	Continuous years (Mean ± SD)	37.57 ± 10.705	
Marital status	Married	52	41.9
	Unmarried	33	26.6
	Divorced	12	9.7
	Widowed	1	0.8
	Cohabitation	26	21.0
Work more than 5 days a week	Always	22	17.7
	Often	31	25.0
	Sometimes	29	23.4
	Rarely	23	18.5
	Never	19	15.3
Work more than 8 h a day	Every day	22	17.7
	Weekly	46	37.1
	Several times a month	29	23.4
	Several times a year	19	15.3
	Never	8	6.5
Average working hours a week (Mean ± SD)		43.78 ± 16.506	

Table 2 Correlation between age, workload and burnout

Items		1	2	3	4	5	6	7	8
1. Age	r_s	1							
	Sig. (2-tailed)								
2. Work more than 5 days a week	r_s	0.176	1						
	Sig. (2-tailed)	0.050							
3. Work more than 8 h a day	r_s	0.066	0.480**	1					
	Sig. (2-tailed)	0.464	**0.000**						
4. Average working hours a week	r_s	−0.210*	−0.451**	−0.498**	1				
	Sig. (2-tailed)	**0.019**	**0.000**	**0.000**					
5. Strain	r_s	−0.253**	−0.072	−0.104	0.190*	1			
	Sig. (2-tailed)	**0.005**	0.425	0.252	**0.035**				
6. Resistance	r_s	−0.108	−0.174	−0.016	0.174	0.416**	1		
	Sig. (2-tailed)	0.234	0.053	0.864	0.053	**0.000**			
7. Exhaustion	r_s	−0.143	−0.117	−0.186*	0.247**	0.661**	0.443**	1	
	Sig. (2-tailed)	0.113	0.197	**0.038**	**0.006**	**0.000**	**0.000**		
8. Level of burnout	r_s	−0.132	−0.185*	−0.141	0.286**	0.754**	0.535**	0.789**	1
	Sig. (2-tailed)	0.143	**0.040**	0.117	**0.001**	**0.000**	**0.000**	**0.000**	

*Correlation is significant at the 0.05 level (2-tailed)
**Correlation is significant at the 0.01 level (2-tailed)

There were weak correlations between the strain and the weekly working hours ($r_s = 0.190$; $p = 0.035$), and moderate correlations between exhaustion, the level of burnout and the weekly working hours ($r_s = 0.247$; $p = 0.006$; $r_s = 0.286$; $p = 0.001$).

The average weekly working hours for all respondents were 43.78 ± 16.51 h (see Table 1). For specialists who have been diagnosed with burnout, the average weekly working time was 48.47 ± 13.72 h. For those who did not receive a positive diagnosis, the average weekly duty time was 37.72 ± 15.22 h.

4 Discussion

The main limitation of this study was the relatively small number of respondents included in the sample, which limits the power of the study. Nevertheless, the results showed that more workload assessed by the number of working hours, resulted in higher levels of burnout, which means that there was a positive correlation between workload and burnout. These findings confirm the conclusions of other studies conducted with a burnout assessment questionnaire different from the Boyko's Inventory [2, 12–15].

Similar to our results Hu et al. found that the workload (number of working hours) has significant association with burnout [16]. Ziaei et al. state that employees who have exposure to excessive workloads will find it difficult to cope with their jobs, which eventually lead to burnout [15].

In conclusion, attention to job burnout should be one of the priorities of those involved in organizational planning. It is important permanently to do diagnosis, assessment and management of burnout related risk factors. In this regard, the identification of workload as a key factor for burnout should provoke urgent actions that minimize the burnout and improve the health of the worker. This can be done by managing human resources well and organized in order to obtain a workforce who is satisfied with his work and thus increase its productivity. Human management departments should be responsible for organizing and planning not only for adequate physical structure, but also for the sufficient number of workers.

Acknowledgements The authors would like to thank the healthcare workers that participated in the study. This work was supported by the Medical University—Plovdiv, Bulgaria under Grant HO-13/2018.

References

1. Pires D, Machado R, Soratto J, Scherer M, Gonçalves AS, Trindade L (2016) Nursing workloads in family health: implications for universal access. Rev Latino-Am Enfermagem 24:e2677
2. Khan F, Rasli A, Yasir M, Khan Q (2019) Interaction effect of social support: The effect of workload on job burnout among universities academicians: case of Pakistan. Int Trans J Eng, Manag, & Appl Sci & Technol 10(13):1–13
3. Dondokambey Z, Saerang D, Pandowo M (2018) The effect of workload and work environment on job burnout (Case Study at Eye Hospital Sulawesi Utara). Jurnal EMBA: Jurnal Riset Ekonomi, Manajemen, Bisnis dan Akuntansi 6(4):3118–3127
4. Kijima S, Tomihara K, Tagawa M (2020) Effect of stress coping ability and working hours on burnout among residents. Med Educ 20:219
5. Luo H, Yang H, Xu X, Yun L, Chen R, Chen Y, Xu L, Liu J, Liu L, Liang H, Zhuang Y, Hong L, Chen L, Yang J, Tang H (2016) Relationship between occupational stress and job burnout among rural-to-urban migrant workers in Dongguan, China: a cross-sectional study. BMJ Open 6(8):e012597
6. López-Núñez M, Rubio-Valdehita S, Diaz-Ramiro E, Aparicio-García M (2020) Psychological capital, workload, and burnout: what's new? the impact of personal accomplishment to promote sustainable working conditions. Sustainability 12(19):8124
7. Maslach C, Leiter M (2016) Understanding the burnout experience: recent research and its implications for psychiatry. World psychiatry 15(2):103–111
8. Sadeghi K, Khezrlou S (2016) The experience of burnout among English language teachers in Iran: self and other determinants. Teach Dev 20(5):631–647

9. Harizanova S, Mateva N, Tarnovska T (2016) Adaptation and vValidation of a burnout inventory in a survey of the staff of a correctional institution in Bulgaria. Folia Med 58 (4):282–288
10. Boiko V (2002) Diagnosis of burnout personality. In Fetiskin NP, Kozlov, Manuilov G (eds) Social and psychological diagnostics of individual and small groups. Institute of Psychotherapy, Moscow, pp 394–399 [In Russian]
11. Boiko V (1999) Methods for diagnosing the level of burnout. In Raigorodskii DY (eds), Practical psycho diagnostics: Methods tests. Samara, pp 161–169 [In Russian]
12. Kelly M, Soles R, Garcia E, Kundu I (2020) Job stress, burnout, work-life balance, well-Being, and job satisfaction among pathology residents and fellows. Am J Clin Pathol 153 (4):449–469
13. Scanlan JN, Still M (2019) Relationships between burnout, turnover intention, job satisfaction, job demands and job resources for mental health personnel in an Australian mental health service. BMC Health Serv Res 19(1):62
14. Starmer A, Frintner M, Freed G (2016) Work–life balance, burnout, and satisfaction of early career pediatricians. Pediatrics 137(4):e20153183
15. Ziaei M, Yarmohammadi H, Moradi M, Khandan M (2015) Level of workload and its relationship with job burnout among administrative staff. Int J Occup Hyg 7(2):53–60
16. Hu N, Chen J, Cheng T (2016) The associations between long working hours, physical inactivity, and burnout. J Occup Environ Med 58(5):514–518

9. [illegible] S, Mateva [illegible] (20[illegible]) Adaptation and validation of a burnout inventory in a survey of the staff of a correctional institution in Bulgaria. Folia Med 58 [illegible]:282–288
10. Boiko V (2002) Diagnosis of burnout [illegible]. In: [illegible] N[illegible] (eds) Stress and psychological [illegible] Institute of [illegible] Moscow, pp [illegible]–359 [illegible]
11. [illegible] (199[illegible]) Methods for diagnosing the level of burnout. In: [illegible] (ed) [illegible] psychodiagnostics [illegible], pp [illegible] [illegible]
12. [illegible] M, [illegible] [illegible]
13. [illegible]
[illegible]

The Advantages of Woven Wraps as Baby Carriers

Irena Šabarić, Stana Kovačević, Franka Karin, and Barbara Šandar

Abstract The paper describes the ergonomic benefits of carrying children in woven wraps. The method of babywearing with woven wraps is a modernized approach which includes several aspects apart from just babywearing. The forms of wearing are proscribed through norms and they stimulate development as well as advance safety and health of the baby. In the experimental part of the paper, a research was conducted into babywearing with woven wraps through a survey among parents aged 18–60 years of age. The purpose of the survey questionnaire was to discover how well informed the participants were as to babywearing with woven wraps and its beneficial effects. How many parents use this method, what are the benefits of this type of babywearing and an overview of ergonomic benefits as well as benefits to the baby's health. The results of the survey prove that the method of babywearing with woven wraps is being recognized by an increasing number of users and it provides numerous benefits for both the child and the parents which is why it is practiced by an increasing number of users.

Keywords Babywearing · Woven wraps · Ergonomic · Body position

---

I. Šabarić · S. Kovačević · F. Karin (✉) · B. Šandar
Faculty of Textile Technology, Textile and Fashion Design, University of Zagreb, Prilaz Baruna Filipovića 28a, 10000 Zagreb, Croatia
e-mail: franka.karin@ttf.unizg.hr

I. Šabarić
e-mail: irena.sabaric@ttf.unizg.hr

S. Kovačević
e-mail: stana.kovacevic@ttf.unizg.hr

B. Šandar
e-mail: bar.sandar@gmail.com

D. Sumpor et al. (eds.), *Proceedings of the 8th International Ergonomics Conference*, Advances in Intelligent Systems and Computing 1313,
https://doi.org/10.1007/978-3-030-66937-9_23

1 Introduction

Ergonomics dictate the advantages and limitations during the product design development through multidisciplinary and interdisciplinary approach. In this way it affects product development and its effective and safe use. Thinking ergonomics during product development results in making people's everyday activities easier in accordance with their behaviour, abilities and limitations [1]. Accordingly, ergonomic approach to babywearing with woven wraps is extremely important as a way of increasing quality of life. Designing products for children and products intended for parents to ease their daily activities presents a great challenge as it is necessary to satisfy a continuous physical, cognitive, emotional and social development [2]. Front or back carriers allow the parent to move and perform activities more easily, while design should also take into account the safety of the child. The head of a newborn as well as young children must be ergonomically supported while components that can break or open and thus endanger the child's safety should be avoided [2]. Babywearing with woven wraps and other types of carriers wrapped around the body can be traced throughout history. Such woven carriers were made from natural threads like cotton, flax and hemp, most often adapted to the traditions, climate, environment and way of life. Traditional carriers around the world are mostly square pieces of fabric. They differ in type of material, length, traditional motifs and tying method, but what they have in common is that they ensure a natural position of the baby's body. The body positions of babywearing are on the chest, on the back or on the hip. The modern approach to babywearing in woven wraps utilises the advantages of modern weaving technology which achieves optimal fabric strength and proportional distribution of the child's body weight in order to reduce the load on the parent's spine. There are more than 50 ways to tie a wrap, each one has its positive effects on the children and the parents [3]. Inspired by the culture of babywearing in softly shaped carriers, Dr. William Sears introduced the term "babywearing" which first appeared in 1978. The term is associated with bonding of a newborn with the mother after birth. Premature children can be worn with a doctor's approval in woven wraps exclusively for the optimal position of the baby's body [4].

2 Properties of Woven Wraps

The width of woven wraps ranges between 45 and 60 cm and they come in 8 standard sizes indicated by numbers 1 through 8. The optimal length of the wrap for the user is the one that allows tying basic front cross carry without any excess fabric. For most people it is size 6 and 4.6 m length. Size 1 indicates a short, woven wrap of 2–2.2 m in length which is used in a traditional hip carry. If metal or wooden rings are sown into a size 1 wrap you get a "ring sling". Sizes 2 and 3 are between 2.6 and 3.1 m long and are used for shorty, hip and back carries. Size 4 is about 3.7 m long and is suitable for all shorty carries, and is also the base size for people of shorter stature. Sizes 5, 6, 7 and 8 are between 4 and 5.7 m long. Sizes 5 and 6 are suitable for people wearing clothes sizes M and L. Wrap sizes 7 and 8 are for people of large and tall build. Woven wraps fabric density is measured in grams per square meter, with the corresponding symbol gms or g/m^2. The weight or the grammage of the fabric is an important characteristic when choosing a wrap as it reveals the thickness of the fabric and the degree of support it provides. The lighter the wrap the thinner the fabric and vice versa. Standard range of woven wraps is between 180 and 260 g/m^2. Wraps weighing between 180 and 220 g/m^2 are considered thin, while ones in the range from 220 to 260 g/m^2 are considered medium, and ones weighing over 260 g/m^2 are considered thick. Wraps above 300 g/m^2 fall in the category of very thick wraps. Grammage, or thickness of the wrap depends on the child's size. Thin wraps are appropriate for small babies and multi-layer carry which ensures adequate support and comfort while babywearing. Wraps of high grammage are adequate for older and heavier children due to better stability in single-layer carry at higher load. The composition of the fabric is another important feature in choosing a woven wrap. Wraps of low grammage containing flax or hemp provide better stability compared to those of higher grammage and made of cotton, wool or silk [5].

2.1 Woven Wraps Tying Methods

A woven wrap can be tied in 50 ways which provide and ideal position for the baby in accordance to ergonomic requirements for all body constitutions. The best known and most commonly used carries are: FWCC front wrap cross carry, FCC front cross carry, cangaroo carry, SHC simple hip carry, Rebozo hip carry, Robins hip carry, Poppins hip carry, HCC hip cross carry, rucksack carry, DH carry "double hammock", shepherd's carry and back wrap cross carry. Front cross carries are most suitable for newborns as they provide the ideal stability for the neck and head. Hip and back carries are ideal for children above 4 months who already have developed a stabile posture as well as head and torso control. It is important that the carry be neat and tight for the safety of the child and the wearer [6].

2.2 Woven Wraps Characteristics

Woven wraps are associated with a terminology common among users which describes particular characteristics and properties. The characteristics "grippy" or "glidey" come in pair and they describe the ease of tightening and layering when tying the wrap. If the wrap is described as very " grippy" it means that it offers more resistance during the tightening of the layers, but if it is very "glidey" that means it slides easily in the layers. The characteristic "saggy" means to sink, and it describes fabric relaxation after a short wearing time when it needs to be re-tightened. “Strechy” means stretchable and it relates to diagonal stretchability of the weave. The “bounce” characteristic is a term which describes the elastic properties of the wrap. “Recoil” describes wraps which, after stretching, return to the original state like a spring. The “diggy” characteristic describes wraps which cut into the shoulders when worn. “Cushy” are wraps which are soft on the shoulders when worn. “Danse” are densely woven wraps while “airy” are breezy woven wraps. The “texture” characteristic describes a distinctive feel of textured wraps while “smooth” describes smooth surface wraps [7].

3 Ergonomic Requirements of Woven Wraps

Babywearing, with newborns in particular, requires ergonomically acceptable body positions to avoid developmental issues. The position of the body is equally important for the parent so woven wraps, ergonomically speaking, have to be sufficiently sturdy, of satisfactory dimensions in form as to evenly distribute the weight of the child and avoid load on the spine of the parent. Positioning infants, especially regarding active neck and back muscles can affect spine development, psychosocial progression and motoric milestone achievements. Carrying babies in arms or in carriers is beneficial for the development of neck muscles and has a positive effect on spine development [8]. The optimal position of a newborn baby which supports the spine is a position in the shape of the letter “C” while knees and hips are in the shape of the letter “M” modelled on Pavlik harness (Fig. 1). Babywearing with a woven wrap or a sling allows for natural position and spine development which newborns take instinctively when held to the chest. Woven wraps allow the optimal position for the parent as well. They adapt to every body type as it is possible to tie them in more than 50 different ways. With each tying method the belt is tied at navel level where the wearer’s centre of gravity is, and it must be tightened well in order to eliminate load on the spine and establish balance. The parent should be aware of the correct posture and determine the optimal hight at which to carry the baby. Wearing the baby to low creates load on the spine, the

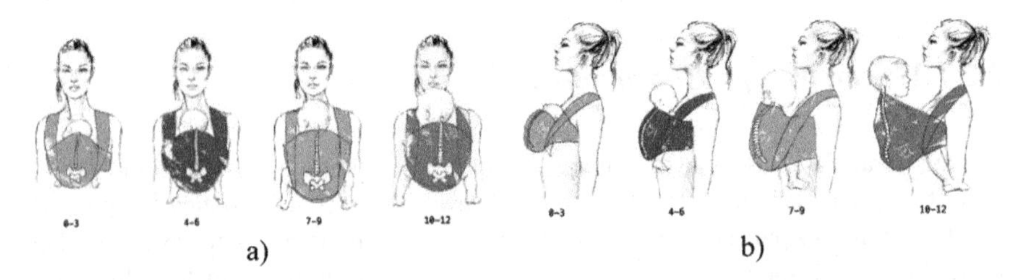

Fig. 1 Proper position of the baby in soft carrier from birth to the first year of age; **a** Front view, **b** Side view; 0–3, 4–6, 7–9, 10–12—months of age of the child

shoulders and the hips and the baby feels heavier. The choice of the tying method depends on parameters such as body type and build, hight, fitness or posture [9]. Ergonomic benefits of babywearing with woven wraps are also beneficial for people with disabilities. For parents with limited mobility and parents in wheelchairs, carrying the child in a woven wrap can enable them to be independent and ease their daily activities with a sense of confidence and independence in raising their child. Wrapping the baby around the body also offers emotional bonding with the child. There are also schools where parents are educated as to all the benefits of babywearing. The content of the courses is based on scientific research in the fields of biology, medicine and psychology. People who have attended the school receive a diploma of a certified counsellor for safe and correct babywearing. One of the best known schools is "Die Trageschule" where teachers are internationally qualified counsellors and health professionals. It has existed since 1998. With its head-quarters in the German city of Dresden.

3.1 The Advantages of Babywearing with Woven Wraps for Children with Difficulties in Hip Development

In the development of newborn motor skills particular attention is given to the proper development of the hips. If the development of the hips is irregular, it requires the application of corrective devices and medical aids like Pavlik harness which belongs to a group of so called dynamic orthosis which enable movement within a limited range [10]. Woven wraps have positive effect on hip disorders due to the correct position of the spine and the "M" position which is equal to the position in the Pavlik harness or the position in a wide swaddle. The said positions have a positive effect on normal hip development. Pavlik recommends active hip movement in order for it to develop normally. The weight of the extremities creates a force which repositions the head of the femur, but the little straps prevent extension and adduction, while allowing safe movement (Fig. 1) [9].

4 Experimental Part

In the experimental part of the paper as part of the research, an anonymous survey was conducted in digital form about the use of woven wraps in babywearing and satisfaction with the use there of. The survey was conducted among parents in the Republic of Croatia in the period from 31.8. to 2.9. 2019. The survey was conducted on a total of 384 people. Participants were both female and male. The survey was divided into three question groups:

1. General information
2. Types of baby carriers
3. Satisfaction and frequency of using woven wraps.

4.1 Survey Results

The first group of questions refers to general information, more precisely to the age of the participant who uses babywearing method with woven wraps and the age of the child being worn in such a carrier. The age of the participants was between 18 and 60 years old. Figure 2 shows that the largest number of participants, 88% between the ages of 26–40 years old, use woven wraps as babywearing method every day. Woven wraps are least common among participants aged 41–60 years old. Figure 3 shows the age of children worn by the participants in woven wraps.

It can be observed that babies aged 12–24 months old are most worn in woven wraps, while newborns and babies up to 6 months old are least worn in woven wraps. The second question group refers to types of baby carriers most commonly used by the participants. Figure 4 shows that 45.7% of the participants use several types of carriers. The method of babywearing with woven wraps is represented in high second place and is increasingly accepted. A small number of participants uses soft structured carriers, and the least number of participants do not practice babywearing.

Third group of questions refers to how satisfied the participants were with woven wraps as they relate to the benefits they provide for children and parents as well as how easy they are to use. Figure 5 shows that the largest number of the participants, 86% rated the babywearing method with woven wraps positively. Negative rating of the woven wraps was given by only 9.1% of the participants, while the least number of particpants have no experience in using woven wraps or have no interest in them.

Figure 6 refers to carry types and ease of tying woven wraps around the body. 27.1% described woven wraps tying method as—easy. In second place came—complicated, which is what 26.3% of the participants said, but they did not give up on this babywearing method because of it. The third place belongs to the rating—good, which is what 20.4% of the participants found while the least number, 7.7% described handling woven wraps as difficult or very difficult.

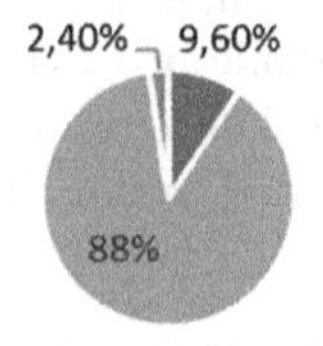

Fig. 2 Participants' age

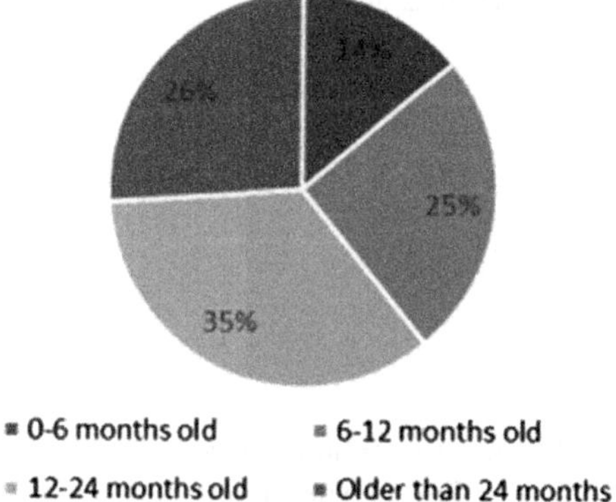

Fig. 3 Children's age

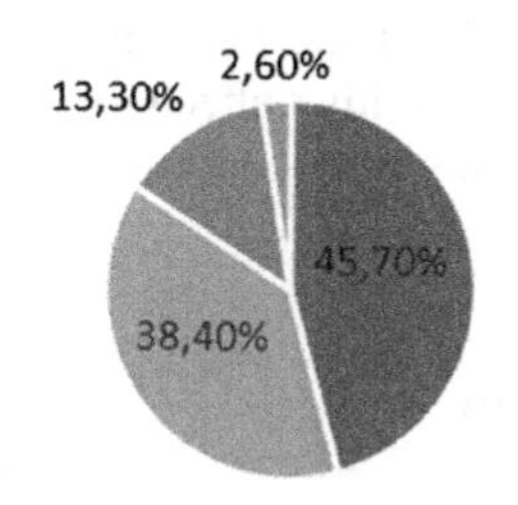

Fig. 4 Types of baby carriers most wraps commonly used by the participants

Fig. 5 Satisfaction with woven

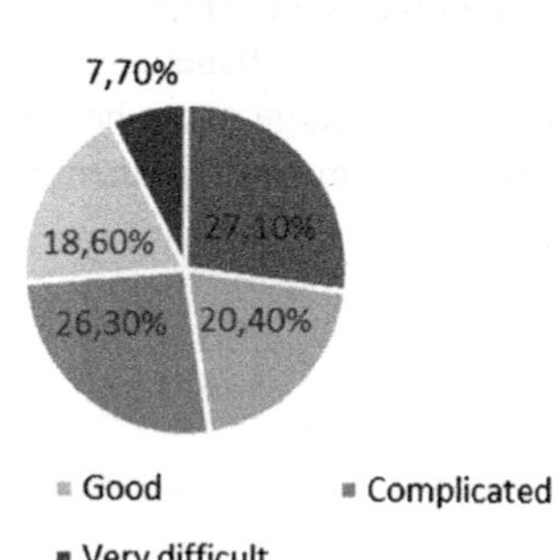

Fig. 6 Rating how easy woven wraps are to tie

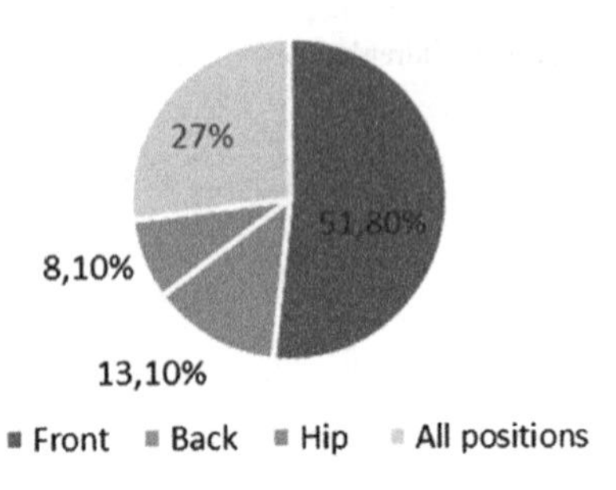

Fig. 7 Babywearing body positions

The most common positions on the body for babywearing in woven wraps are shown in Fig. 7. The largest number of participants 51.8% carry the baby in the front to control the baby's needs. All body positions are practiced by 27% of the participants. 13.1% of the participants carry the baby on the back, while 8.1% choose hip carry. Positive effects of babywearing method with woven wraps on children and parents were confirmed by 88.3% of the participants [11].

5 Conclusion

Babywearing in soft carriers is a method that has been in use for centuries up until today. There are numerous benefits of babywearing with woven wraps recognized by a large number of modern parents and is increasingly practiced around the world. It satisfies ergonomic requirements by providing the optimal body position of the baby which positively affects further spine and motor skills development and prevents developmental deformities. In addition to the positive effect on the development, babywearing with woven wraps also helps with developmental disorders in newborns, particularly hip problems. For such disorders, woven wraps and the numerous ways they can be tied, allow for body positions intended for normal hip development. Advance technologies and innovations have shortened the production time for woven wraps and increased the properties related to easier fabric breaking which allows for easier handling and tying of woven wraps. Another advantage is the wraps' adaptability to different body types of the wearer. Research shows that the benefits of babywearing with woven wraps are well accepted by users on a large scale because it combines tradition, technological advancements and ergonomic requirements which has a positive effect on both the children and the parents.

References

1. Muftić O, Veljović F, Jurčević Lulić T, Milčić D (2001) Basics of ergonomics. University of Sarajevo, Sarajevo
2. Jurčević Lulić T (2015) Ergonomics and children's safety. Paediatr Croat, Suppl 59, pp 155–160
3. Van Hout IC (2011) Beloved burden—Babywearing around the world. KIT Publishers, Amsterdam, Nizozemska
4. Blois M (2005) Babywearing, the benefits and beauty of this ancient tradition. Pharmasoft Publishing L.P., TX, USA
5. Simonović S (2020) What is weave density and what does gsm mean? https://nosi.me/2017/09/gustina-tkanja-gsm-oznaka/. Last accessed 22 September
6. Božić I (2016) Some new old doubts about the woven scarf - single-layer and multi-layer embroidery. https://nosenjebeba.com/2016/01/18/neke-nove-stare-dvojbe-u-vezi-tkane-marame-jednoslojni-i-viseslojni-vezovi/. Last accessed 22 September 2020
7. Simonović S (2018) Characteristics of woven scarves. https://nosi.me/2018/03/terminologija-karakteristike-tkanih-marama/. Last accessed September 22
8. Safeer F, Siddicky DB, Bumpass Akshay K, Stewart A, Tackett RE, McCarthy E, Mannen M (2020) Positioning and baby devices impact infant spinal muscle activity. J Biomech 104:10974
9. Hsu JD, Michael JW, Fisku JR (2008) Atlas of orthoses and assistive devices, 4th edn. American Academy of Orthopaedic Surgeons Mosby Elsevier, Philadelphia, 35:464–477
10. Cassidy JT, Petty RE, Laxer RM, Lindsley CB (2010) Textbook of pediatric rheumatology, 6th edn. Saunders/Elsevier, Philadelphia
11. Šandar B (2019) Woven wraps for children. University of Zagreb Faculty of Texstile Tehnology, Zagreb, urn:nbn:hr:201:279632

The Moisture Management Properties of Laundered Hospital Textiles

Anita Tarbuk, Tihana Dekanić, Sandra Flinčec Grgac, and Ivana Čorak

Abstract The influence of repeated laundering to the change in moisture management properties of hospital protective textiles was researched in this paper. For that purpose, green cotton fabric for surgical gowns and white one for linen were subjected to repeated laundering according to ISO 15797:2017 up to 25 washing cycles. *Washing procedures for white workwear and/or sensitive coloured trimmings—Peracetic acid (PAA) bleach* were performed in Wascator FOM71 CLS machine at 75 °C using Reference Detergent A1 with optical brightener by WFK. The physical-chemical properties were monitored on FTIR-ATR, Spectrum 100, PerkinElmer after 1st, 3rd and 25th cycle. The liquid moisture management properties were determined according to AATCC TM 195-2017 on Moisture Management Tester (MMT M290, SDL Atlas). It has been showed that hospital protective textiles contain finishing treatment, which removes during laundering, even after one washing cycle, changing the moisture management properties. Fabrics goes from water penetration, through moisture management to fast absorbing and quick drying fabrics. The MMT results indicate better comfort of such treated fabrics.

Keywords Cotton · Repeated laundering · Peracetic bleach · FTIR · MMT

1 Introduction

There are lots of requirements on textiles that are aimed for use in hospital environment, especially non-disposal ones, such as good antimicrobial and mechanical properties, minimal textile dust generation, comfort [1–3]. After laundering process, the cleanliness is important, but the aspect of hygiene is a primary criterion [4]. Different impurities and microorganisms from contaminated textiles in hospitals, if

A. Tarbuk (✉) · T. Dekanić · S. F. Grgac · I. Čorak
Department of Textile Chemistry and Ecology, University of Zagreb
Faculty of Textile Technology, Zagreb, Croatia
e-mail: anita.tarbuk@ttf.unizg.hr

D. Sumpor et al. (eds.), *Proceedings of the 8th International Ergonomics Conference*, Advances in Intelligent Systems and Computing 1313,
https://doi.org/10.1007/978-3-030-66937-9_24

not adequately disinfected, can become one of the possible carriers of hospital infections as well as a source of hospital pathogens. Laundering, e.g. washing the laundry, is a complex process of soil and/or stain removal that occurs in an aqueous medium. During that process, four parameters influence the final results—mechanical agitation, chemistry, temperature and time. According to Sinner, if one of these parameters is increasing, the other ones must decrease, resulting in different washing effect [1–6]. Therefore, for laundering hospital textiles, a satisfactory disinfecting effect is also required what can be achieved by combination of chemistry and temperature. The composition of detergent, as well as its combination with different additives, like bleaches and disinfection agents, including temperature, can be relevant for process efficiency. Peracetic acid (PAA) is often used in hospital environment—for disinfecting instruments due to its virucidal efficacy [7], as powerful oxidant which acts against spoilage and pathogenic microorganisms, and lately as oxidative bleaching and disinfection agent in the industrial laundering, especially for hospital textiles [8, 9]. Additionally, PAA is eco-friendly due to the decomposition of PAA to hydrogen peroxide and acetic acid. Among numerous requirements on hospital textiles, for gowns and linen, a thermophysical and thermophysiological comfort is especially important. For gowns, a fabric must allow air, and especially the moisture (water vapor) generated by the exudation of perspiration from the skin during physical activity, to pass through the fabric [3, 10–12]. For lining, it has to be fast drying fabric with the ability to take away liquid by capillary force. The moisture management properties are one of the key performance criteria regarding the comfort level of fabric and depends on numerous factors, such as fabric composition and structure, surface modification, finishes etc.

Repeated laundering causes oxidative damages of cotton fabric, changing fabric whiteness and color fastness, mechanical and surface properties. For that reason, the change in physical-chemical properties as well as moisture management properties of laundered hospital textiles were researched in this paper.

2 Materials and Methods

For the research of the influence of repeated washing on hospital protective cotton fabric properties, two commercial cotton fabrics were selected, green fabric (G) for surgical gowns colored with vat dye, and optical brightened white fabric (B) for hospital linen. Both fabrics have mass per unit area 178 g/m^2, yarn fineness 35 tex, yarn density of warp: 24 cm^{-1} and weft: 21 cm^{-1}. Fabrics were subjected to repeated washing (25 cycles) according to ISO 15797:2017 *Textiles—Industrial washing and finishing procedures for testing of workwear: Washing procedures for white workwear and/or sensitive coloured trimmings—Peracetic acid bleach* in Wascator FOM71 CLS machine (Electrolux). 4 g/l of Reference Detergent A1 with optical brightener (WFK) with addition of 2 g/l of Proxitane 523 (Ivero), a peroxyacetic acid—mixture of 5% acetic acid and 20% H_2O_2, were applied at 75 °C.

Fabric properties were determined before and after 1st, 3rd and 25th washing cycle. The physical-chemical properties were monitored on FTIR-ATR Spectrum 100, Perkin Elmer. 16 scans at a resolution of 4 cm^{-1} were recorded for each sample between 4000 and 380 cm^{-1}. The liquid moisture management properties were determined according to AATCC TM 195-2017 *Liquid Moisture Management Properties of Textile Fabrics* on Moisture Management Tester (MMT M290 by SDL Atlas).

3 Results and Discussion

Repeated laundering has high influence to textile properties. Therefore, its influence to the change in moisture management properties of hospital protective textiles was researched in this paper. Green cotton fabric (G) for surgical gowns and optically brightened white one (B) for linen were subjected to repeated laundering. The physical-chemical properties after 1st, 3rd and 25th cycle were monitored on FTIR-ATR, and compared to unwashed fabric (0). Results are presented in Fig. 1.

In Fig. 1 characteristic peaks for cotton can be seen. On both unwashed fabrics (0), a peak at the wave number 2849 cm^{-1} can be observed, attributed to the stretching within the CH_2 group. This peak is not visible on the washed fabric samples, what can be an indication of finishing based on waxes on surface. In Fig. 1a. a peak at 1573 cm^{-1} which indicates vibrations within the COO– groups is visible on unwashed white fabric (B0), whilst after 1st washing cycle no longer appears. A peak of very low intensity at 890 cm^{-1} which is formed due to symmetrical stretching within the C(1)–O–(4)C bond can be seen in sample B0. After washing, the intensity of the peak in this area is higher, which indicates the removal of substances that interfered with the detection of the resulting vibrations present in the sample. In Fig. 1b. additional peak appears at wave length 1728 cm^{-1}. This peak is attributed to C=O bond from vat dye, which contains two or more keto groups. To be applied on cotton cellulose, it has to be vatted to its alkali-soluble enolic (leuco) form. Afterwards, reoxidation occurs resulting in excellent fastness. After 25 washing cycles, fabric is slightly damaged, so the peak disappears. At the same time, a peak not noticed in unwashed green fabric (G0) at 1573 cm^{-1} attributed to the vibrations within the COO– groups is getting visible.

Results of the liquid moisture management properties determined according to AATCC TM 195-2017 on MMT M290 are presented in Tables 1 and 2. The overall grading in regard to fabric performance is described automatically as "Type of Fabric" and listed in tables as well. The wetting time (WT) measured on MMT is the time period in which the top and bottom surfaces of the fabric just start to get wetted, and corresponds to the absorbency drop test. From the results shown in Table 1 it can be seen that white cotton fabric (B0) shows maximum wetting time of 120 s, suggest hydrophobic treatment, what corresponds FTIR-ATR. Wetted radius of 5 mm, slow spreading speed, small spreading area, and very good one-way transport suggests that fabric B0 is characterized as water penetration

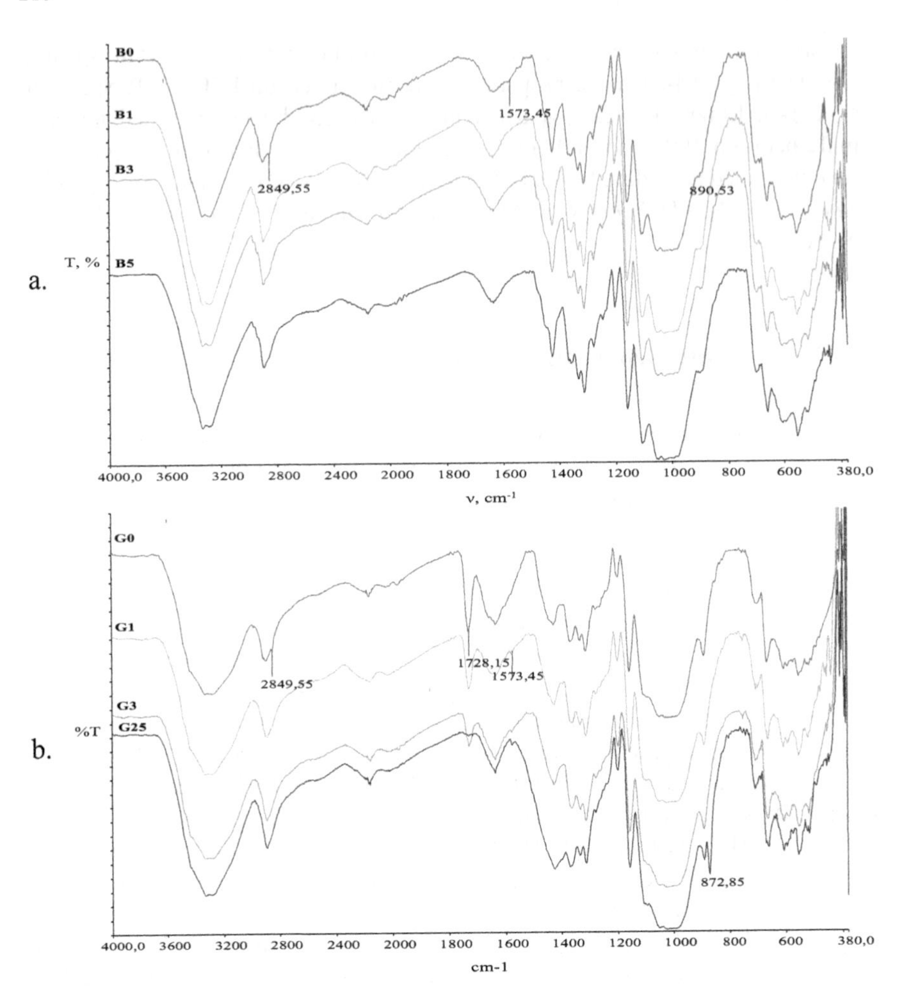

Fig. 1 FTIR-ATR of hospital cotton fabric: **a** white (B) and **b** green (G) before (0) and after 1st (1), 3rd (3) and 25th (25) laundering cycle

fabrics, indicating good repellency but having water vapor perspiration. After 1st washing cycle (B1) wetting time is lower. The reason for that is that the finishing agent was partially washed off, confirmed by FTIR-ATR. However, small amount of finishing agent is still present, so the fabric B1 is also characterized as water penetration fabric. After 3rd washing cycle most of the finishing agent is washed off, and surface changed. Absorption rate, which represents the average speed of liquid moisture absorption for the top and bottom surfaces during the initial change of water content during a test, is getting higher, indicating better capillarity.

For that reason, the spreading speed, which represents the accumulated rate of surface wetting from the center of the specimen where the test solution is dropped to the maximum wetted radius, is faster (from 0 mm/s for B0 to 3.5 mm/s for B3).

Table 1 Moisture management properties of white fabric (B) before and after laundering

Fabric		B0		B1		B3		B25	
		Mean	*CV	Mean	*CV	Mean	*CV	Mean	*CV
WT (s)	T	120	0	36.88	1.189	3.13	0.034	2.9	0.032
	B	6.94	0.108	6.21	0.356	3.26	0.031	3.07	0.025
AR (%/s)	T	0	0	14.38	0.633	64.79	0.087	67.4	0.014
	B	60.59	0.093	48.7	0.54	66.96	0.037	60.62	0.011
MWR [mm]	T	0	0	10	0.5	20	0	22	0.12
	B	5	0	10	0.5	20	0	23	0.12
SS (mm/s)	T	0	0	0.64	0.741	3.59	0.034	4.48	0.041
	B	0.7	0.1	0.96	0.256	3.46	0.045	4.38	0.054
R (%)		927.85	0.051	635.52	0.346	−12.25	1.227	6.85	1.146
OMMC		0.6405	0.0246	0.6137	0.1248	0.4053	0.0676	0.4538	0.0211
Type of fabric		*Water Penetration Fabric*		*Water Penetration Fabric*		*Fast Absorbing and Quick Drying Fabric*		*Fast Absorbing and Quick Drying Fabric*	

Wetting Time—WT, Absorption rate—AR, Maximum wetted radius—MWR, Spreading speed—SS, top surface—T, bottom surface—B, Accumulative One-way Transport Capability—R, Overall (liquid) Moisture Management Capability—OMMC, *CV—coefficient of variation

Table 2 Moisture management properties of green fabric (G) before and after laundering

Fabric		G0		G1		G3		G25	
		Mean	*CV	Mean	*CV	Mean	*CV	Mean	*CV
WT [s]	T	4.23	0.207	5.88	0.223	3.95	0.305	2.56	0.02
	B	4.72	0.215	5.71	0.229	3.89	0.331	2.67	0.019
AR (%/s)	T	32.97	0.293	38.13	0.238	77.3	0.112	76.17	0.004
	B	55.17	0.144	58.03	0.17	75.13	0.11	68.13	0.006
MWR (mm)	T	27	0.101	20	0	24	0.093	26	0.086
	B	27	0.101	22	0.124	24	0.093	25	0
SS (mm/s)	T	3.42	0.162	2.43	0.168	4.47	0.084	5.68	0.019
	B	3.26	0.155	2.55	0.173	4.53	0.109	5.48	0.005
R (%)		121.87	0.76	179.68	0.446	41.09	0.513	−8.55	0.515
OMMC		0.5045	0.4782	0.4607	0.5006	0.5827	0.5053	0.5045	0.4782
Type of fabric		*Fast Absorbing and Quick Drying Fabric*		*Moisture Management Fabric*		*Fast Absorbing and Quick Drying Fabric*		*Fast Absorbing and Quick Drying Fabric*	

Wetting Time—WT, Absorption rate—AR, Maximum wetted radius—MWR, Spreading speed—SS, top surface—T, bottom surface—B, Accumulative One-way Transport Capability—R, Overall (liquid) Moisture Management Capability—OMMC, *CV—coefficient of variation

After 25th cycle, SS is faster (4.4 mm/s) and AR higher. Therefore, white fabrics B3 and B25 are characterized as Fast Absorbing and Quick Drying Fabric. Accumulative one-way transport capability (R) represents the difference between the area of the liquid moisture content curves of the top and bottom surfaces of a specimen with respect to time. For fabric B0 is 927.85% indicating that water content on top surface is much higher than on bottom. B3 show negative values indicating that water content top surface is lower than the one on fabric bottom surface, suggesting that the liquid introduced to the bottom surface transfers to the top surface slowly. Overall (liquid) moisture management capability (OMMC) is calculated by combining three measured attributes of performance: the liquid moisture absorption rate on the bottom surface, the one-way liquid transport capability, and the maximum liquid moisture spreading speed on the bottom surface. It represents an index of the overall capability of a fabric to transport liquid moisture.

For the difference of white fabric, hospital green fabric G0, due to dyestuff blocking the active groups have smaller absorption, and it is already characterized as Fast absorbing and quick drying fabric. During 1st laundering cycle it one-way transport changed from fair to very good, so the fabric G1 is characterized as moisture management fabric. Further laundering resulted in Fast absorbing and quick drying fabric, probably because fabric was step by step worn out, and absorption was faster and higher.

4 Conclusion

It has been showed that hospital protective textiles contain finishing treatment, which removes during laundering, even after one washing cycle, changing the moisture management properties. Fabrics goes from water penetration, through moisture management to fast absorbing and quick drying fabrics, what indicates better comfort of such treated fabrics.

Acknowledgements The work has been supported by Croatian Science Foundation under the project UIP-2017-05-8780 HPROTEX. The authors thank company "DM TEKSTIL KROJAČKI OBRT" Ozalj for donation of hospital protective cotton fabrics.

References

1. Budimir A, Bischof Vukušić S, Flinčec Grgac S (2011) Study of antimicrobial properties of cotton medical textiles treated with citric acid and dried/cured by microwaves. Cellulose 19 (1):289–296
2. Gao Y, Cranston R (2008) Recent Advances in Antimicrobial Treatments of Textiles. Text Res J 78(1):60–72

3. Tarbuk A, Flinčec Grgac S, Dekanić T (2019) Wetting and wicking of hospital protective textiles. Adv Technol 8(2):5–15
4. Neral B, Šostar-Turk S, Fijan S (2011) Environmental impacts of various disinfection procedures during laundering. Tekstilec 54(7–9):149–171
5. Zoller U (1999) Part A: properties, disinfection and preservation in detergents. In: Handbook of detergents, Marcel Dekker, Inc., Basel
6. Pušić T, Dekanić T, Orešković M, Soljačić I (2017) Tehnološka unapređenja u industrijskim praonicama rublja. Annual 2016 of the Croatian Academy of Engineering. 23, Zagreb (2017)
7. Rodgers SL, Cash JN, Siddiq M, Ryser ET (2004) A comparison of different chemical sanitizers for inactivating Escherichia coli O157: H7 and Listeria monocytogenes in solution and on apples. lettuce, strawberries, and cantaloupe. J Food Prot 67(4): 721–731
8. Kitis M (2004) Disinfection of wastewater with peracetic acid: a review. Environ Int 30 (1):47–55
9. Forte Tavčer P, Križman P (2003) Bleaching of cotton fibers for sanitary products with peracetic acid (In Croatian: Bijeljenje pamuka za sanitetske proizvode s peroctenom kiselinom). Tekstil 52(7):309–315
10. Tang K-P, Kan C-W, Fan J (2014) Evaluation of water absorption and transport property of fabrics. TextE Prog 46(1):1–132
11. Das B, Das A, Kothari VK, Fanguiero R, De Araújo M (2007) Moisture Transmission through Textiles, Part I: Processes involved in moisture transmission. AUTEX Res J 7 (2):100–110
12. McQueen RH, Batcheller JC, Mah T, Hooper PM (2013) Development of a protocol to assess fabric suitability for testing liquid moisture transport properties. J TextE Inst 104(8):900–905

Research Methodology Used to Investigate the Effects of Noise on Overhead Crane Operator's Performances

M. Masullo, R. A. Toma, A. Pascale, G. Ruggiero, and L. Maffei

Abstract Cranes are among the most hazardous mechanical equipment and are widely used in several industrial fields. During heavy loads' handling, crane operators are exposed to high risks of accidents. Environmental and psychological factors can detriment crane operators' efficiency. To investigate the effect of noise on overhead crane operators' performances under mental fatigue, a Virtual Reality overhead crane simulator has been built to be used in experimental sessions in laboratory. This paper illustrates the research methodology of an ongoing study. In the first part is discussed the approach used to investigate the issues of crane operation safety (risk factors, accidents, errors, training, workers wellbeing), in the second one is described the VR simulator and the experimental design used for the pilot study on a group of inexperienced operators. In the last part are showed the methods used to obtain and analyze the data and discussed the limitations and the design of corrective measures of the experimental environment.

Keywords Crane · Safety at work · Virtual reality · Occupational noise · Mental fatigue

M. Masullo · R. A. Toma (✉) · A. Pascale · L. Maffei
Department of Architecture and Industrial Design, Università degli Studi della Campania "Luigi Vanvitelli", 81031 Aversa, Italy
e-mail: roxanaadina.toma@unicampania.it

M. Masullo
e-mail: massimiliano.masullo@unicampania.it

G. Ruggiero
Department of Psychology, Università degli Studi della Campania "Luigi Vanvitelli", 81100, Caserta, Italy
e-mail: gennaro.ruggiero@unicampania.it

D. Sumpor et al. (eds.), *Proceedings of the 8th International Ergonomics Conference*, Advances in Intelligent Systems and Computing 1313,
https://doi.org/10.1007/978-3-030-66937-9_25

1 Introduction

In many industrial settings, crane operations are very high-risk tasks. As reported by Milazzo et al. [1] the fatalities were caused by overhead power line electrocution (25%), crane collapse (14%), falls, (9%) or are associated with crane operators being struck by crane loads (21%), struck by crane booms/jibs (12%), struck by cranes or crane parts (7%), caught in/between (5%), or other causes (7%). Environmental work conditions (unstable ground conditions, dangerous weather, poor light conditions and high noise levels) [2] and psychological factors [3] can detriment crane operator's efficiency especially when the handling of loads is performed by inexperienced or inattentive operators. Due to the high number or accidents occurring all over the world [1–4] the governments are investing in crane operator's knowledge and training. Virtual Reality tools allow crane operators to gain skills and expertise in scenarios that replicate real worksite conditions [5] as well as to induce stress in laboratory to investigate decision making and the physiological response during stressful events in order to reduce the probability of errors [6].

High noise levels and noise types can act as stressors and negatively influence worker's wellbeing, their physiological response of the central and autonomic nervous system as well as their cognitive functions [7–11].

To investigate the effect of noise on overhead crane operators' performances, under mental fatigue, a Virtual Reality overhead crane simulator has been built to be used in experimental sessions in laboratory. This study illustrates step by step the methodology used and the corrections which were introduced to improve the experimental design of a so complex experiment.

2 Methodology

2.1 Investigation on Crane Operation Safety Issue

As first step a bibliographic research has been carried out to investigate the main crane operation safety issues and, studying the Health and Safety at Work Regulations, identify the correct procedure of using cranes: Personal Protective Equipment (PPE), functional checks before use, maneuver actions, tips and warnings necessary to operate safely, prohibited maneuvers, functional checks after the use and at the end of the shift.

After that the most hazardous scenarios in Italy have been investigated through the web-tool for the qualitative updating of cases of fatal accidents Infor.Mo (last access 18.10.2019) belonging to the Italian National Institute for Insurance against Accidents at Work (INAIL). It allows to investigate in a standardized way the accident dynamics. For each accident is possible to analyze the building blocks of an injury (accident, contact, damage) and identify the risk factors (determinants) that contribute to the accidental event and any other factors (modulators) able to prevent, mitigate or also worsening the resulting biological damage [12].

Typing the keyword "*carroponte*" ("*overhead crane*" in English) a total of 42 injuries were detected for serious cases section and 70 for fatal cases. The cases identified were collected in a database and the leading causes of crane accidents and incidents were analyzed. The causes of the identified accidents are due to one or more errors related to incorrect use of the bridge crane, incorrect management of materials, environmental characteristics, incorrect or non-use of Personal Protective Equipment (PPE), wrong actions or negligence by the injured party or by third parties. Moreover, the principal common injuries have been found to be amputations, fractures and crushing of the chest or skull which in the worst cases can lead to death and at best to the loss of many days of work (up to 400 days). All the workers involved were male except in one case and 29% were foreign workers.

2.2 Virtual Reality Simulator

Visual virtual Environment. The Virtual environment is representative of an industrial workspace of about 580 m^2 equipped with storage units and a single girder overhead crane. The space is divided into three different areas: (1) a loading area where the operator can hook the load, (2) a transit area where the load has to be safely handled avoiding the central obstacle with a variable position, (3) a target area where the operator has to place the load (Fig. 1).

To build the virtual environment was used ArchiCAD for the building of the 3D Model and 3D Studio Max to model and optimize the crane and other objects. The visual aspect of the model (surface materials and lighting) has been implemented and optimized in UE4, as well as the overhead crane controls have been coded. The crane has a field of action equal to half of the building surface. It can simulate the

Fig. 1 Virtual reality crane simulator

movements of the beam along the tracks, the transversal movement of the trolley, and the up and down movement of the hook, as in reality.

The Oculus Rift S was chosen as virtual reality exploration system and the related controllers were used for the three movements of the crane allowing the operator to calibrate the speed of the movements. At the end, an application was created, automating the experimental session.

Audio Materials. As background noises the soundtracks of three different industrial noises were used: a low frequency noise (LF), a high frequency noise (HF) and a modulated noise (MOD) fixed at 80 dB(A) (±1 dB). An additional soundtrack with low background noise fixed at 51 dB(A) was used as control condition (CTRL). To enhance the realism of the virtual scenarios the soundtracks of the beam, the trolley and the hook were added and synchronized with the crane movements and they were set with the help of an expert crane operator. The Sennheiser HD 280 Pro headphones were used to playback the sounds stimuli. The audio playback system has been calibrated by means of a Mk1-Cortex manikin, a Symphonie card 01 dB and a laptop (Fig. 3a). The A-weighted and linear sound equivalent levels with the 95th and 5th percentiles and spectrogram of the soundtracks were extracted using the Artemis Suite software.

Cognitive tasks. For this study three different task were selected: the Rey Test, the Verbal Fluency Test, and the Backward Counting test.

During the Rey Test (REY) subjects were asked to recall as many words as possible of a presented list of 15 words in order the evaluate the verbal memory. For this experiment 10 lists were used.

During the Verbal Fluency test (VF) subjects were asked to list as many words as possible with a target letter to evaluate the ability to recall long-term phonemic memory. For the experiment 10 letters (S, C, A, P, I, M, D, T, B, F) were selected.

During the Backward Counting test (BC) subjects were asked to count down by seven from a target number with the aim of evaluates the executive functions. For the experiment 10 target number (90, 83, 76, 68, 96, 81, 72, 64, 93, 84) were chosen.

For the training session a different list of words, target letter and target number were used.

2.3 Training Materials

Two training sessions have been prepared for the study, a traditional one with slides and another one in virtual reality.

For the first one, the content for the training was selected based on the health and safety at work regulations. The lesson, lasting about 10 min, focused on the description of the different types of cranes and their use; on the main risks associated with the use of cranes and on the correct procedure of use. Moreover, advices and warnings were given and the prohibited maneuvers were highlighted.

The second one, is a VR session training which allowed the subjects to learn how to use the virtual reality crane simulator and to practice the selected cognitive tasks before starting the experiment.

2.4 Physiological Parameters

The physiological parameters were observed by means the Equivital (ADInstruments) wireless physiological monitoring system. The system is composed by a Sensor Electronics Module (SEM), a set of Sensor Belts with different sizes and ancillary devices. It can record ECG data and heart rate, respiration rate, skin temperature, Galvanic Skin Resistance and body orientation, motion, and activity.

Before starting the experiment, the Sensor Electronics Module (SEM) was configured using the Equivital Manager Software. The clinical mode for Heart Rate and the frequency 16 Hz were chosen.

2.5 Experimental Design

A within subject design was prepared. The four background noises and the cognitive tasks were presented to the subjects in a Latin Square Balanced Sequences. The experimental protocols included the two training sessions above explained, a Go/No Go test session administered to assess inhibitory control and the sensitive interference, and the maneuver test of the overhead crane in VR (Fig. 2). During the experiment was involved a preliminary group of 18 inexperienced operators.

2.6 Test Session Procedure

The researcher explained the experimental design to the subject and before starting the test session had asked him to sign the data processing and the informed consent for the use of Equivital belt in order to monitor the Heart Rate, Galvanic Skin Response, Respiration Rate and Skin Temperature. Then to each subject was asked to participate in the traditional training session previously explained and if needed the researcher gave additional information and answered all questions. After that, a Go/No go test session lasting about 5 min was administered. Then the subject was asked to wear the head mounted display and the headphones and take part in virtual reality training session how many times he wanted. When the subject was ready, the researcher started the test session (Fig. 3b).

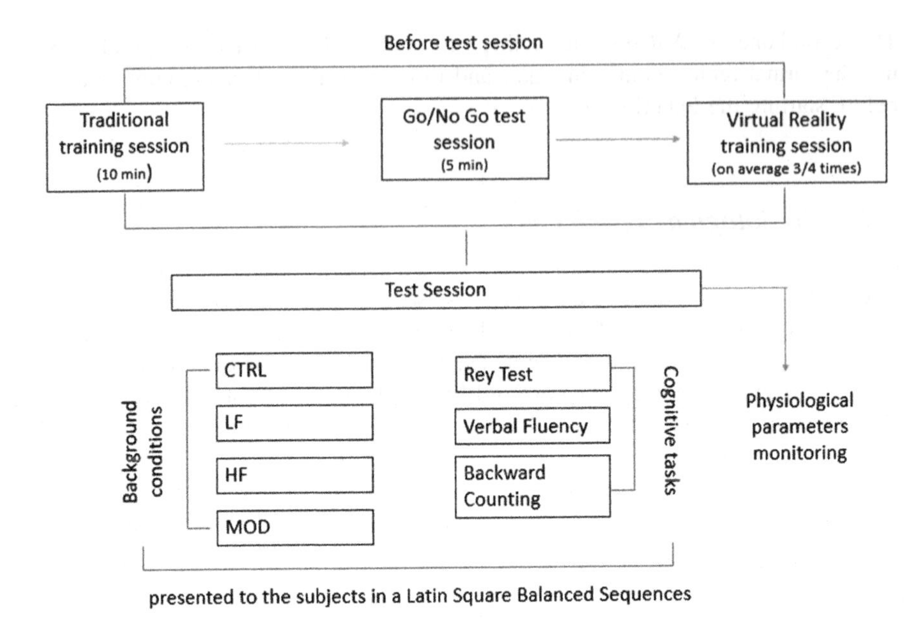

Fig. 2 Phases of the experimental design

Fig. 3 **a** Calibration of the audio playback system; **b** test session involving a preliminary group of inexperienced operators

During the test, the subject was asked to use the crane to move the target load from a starting point of the virtual industrial building to an end point avoiding the obstacle placed in the transit area and in the meantime perform the three cognitive tasks.

With regard the Rey Test the list of 15 words was visually presented for 42 s, within this time frame the subject was asked to focus on memorizing the presented list and after that the subject had to immediately recall as many words as possible for other 30 s. In the Verbal Fluency task, the subject was asked to list as many words as possible with the target letter visually presented. In the Backward Counting test, the subject was asked to count down by sevens from the target number appeared in VR scenario. Verbal Fluency test and Backward counting test were administered for 30 s. On the right the subject could visualize a traffic light. When it was red the subject was asked to focus on task presented (list of words, target letter or number) and when the light turned green the subject was asked to carry out the task.

The test room temperature of the SENS i-Lab was kept between 20–24 °C, as well as the relative humidity at 50% and the air flow rate at 0.5 m/s.

3 Data Acquisition and Analysis

The Virtual Reality simulator for each scenario record every 200 ms the following parameters: (1) time; (2) load/hook coordinates (x, y, z); (3) trolley coordinates (x, y, z); (4) presence or absence of the operator within the danger area (r = 50 cm) around the load (true/false); (5) speed (m/s); (6) Presence or not of simultaneous maneuvers of the load along two axes (true/false). These parameters allow to monitor some of the most common and dangerous errors during the handling of heavy loads operations and in particular: (1) Oblique maneuver, (2) proximity of crane operator to the load, (3) excessive oscillation of the load, (4) maneuver speed and (5) maneuver of the load with irregular movements. The total time spent by operator committing errors was considered and the phases in which cognitive tasks were performed were excluded. Five one-way ANOVAs were carried out to investigate on the dependency between the typology of industrial noise and the errors during overhead crane maneuver.

The physiological data were acquired using the LabChart Software, which allowed experimenter to monitor in real time the principal physiological parameters and record all the data. Heart Rate and Galvanic Skin Resistance of the preliminary experimental session were analyzed for LF, HF and MOD scenarios every 30 s and compared with CTRL scenario. Four one-way ANOVA were carried out on the variables: GSR, average RR interval and LF/HF ratio.

The data of the cognitive tasks were also collected during the test session by the experimenter using a specific datasheet.

4 Preliminary Results and Corrective Measures

Data analysis of the test shows that, among the errors, the oblique manoeuvres were negatively influenced by the noise. Excessive oscillations varied slightly, while the remaining variables did not change. HF noise conditions seems to be the worst.

On the other side, the results of the electrodermal activity, as well as the ratio of low frequency power to high frequency power of the heart rate, show that the typology of noise influenced these variables. In particular, the modulated noise seems to induce the highest increments of GSR and of the LF/HF ratio, when compared with the control condition. Less evident is the differences between at the HF and LF noise exposure. The analysis of cognitive data pointed out that almost all the subjects continued handling the load only after completing the cognitive tasks. This vanished the aim to assess the cognitive performance during dual tasks.

Given the complexity of the experimental design and the results obtained, some corrective measures have been designed. To increase the output reliability, the control condition has been fixed at the beginning of each test session. This should help to have more stable values of the physiological variables to which refer the changes.

Additionally, to make loads handling operation more complex, the maneuver area has been reduced adding more storage units and obstacles. This should make more evident some errors (e.g. excessive oscillations).

Corrections also concerns the cognitive tasks, with the aim to avoid the subject to interrupt the manoeuvres when perform these tasks.

5 Conclusions

Using a Virtual Reality Overhead Crane Simulator, the effects of noise on overhead crane operators' performances, under mental fatigue were investigated by mean of preliminary test. The maneuver test and the physiological data of the LF, HF and MOD scenarios were compared with the CTRL scenario. Main results showed that: Oblique maneuver, Galvanic Skin Resistance and LF/HF ratio of Heart Rate changes as the exposure to noise changes.

Preliminary results showed that the effects of noise on the other typologies of maneuver errors were not yet evident. At the same time, also the visual presentation (at screen) of cognitive tasks has avoided the participants to perform the cognitive tasks as dual tasks. To increase the output reliability some corrective measures were designed.

References

1. Milazzo MF, Ancione G, Spasojevic Brkic V, Vališ D (2017) Investigation of crane operation safety by analyzing main accident causes. In: Walls L, Revie M, Bedford T (eds) Risk, reliability and safety: innovating theory and practice. Taylor & Francis Group, London
2. Small entity compliance guide for the final rule for cranes and derricks in construction (2014) U.S. Department of Labor Occupational Safety and Health Administration, OSHA3433-10R
3. Tam VWY, Fung IWH (2011) Tower crane safety in the construction industry: a Hong Kong study. Saf Sci 49:208–215
4. Skiba R (2020) Best practice standards and methodology for crane operator training—a global perspective. J Transp Technol Sci Res Publishing 10:265–279

5. Dhalmahapatra K, Pradhan O, Das S, Singh K, Maiti J (2018) Prioritization of human errors in EOT crane operations and its visualization using virtual simulation. In: 4th International conference on recent advances in information technology. IEEE, Dhandad, India
6. Mosquera-Dussán O, Guzmán-Pérez D, Terán-Ortega P, García J, Trujillo-Rojas C, Zamudio-Palacio J, Botero-Rosas D (2020) Decision making, stress assessed by physiological response and virtual reality stimuli. Revista Colombiana de Psicología 29(2):89–103
7. Pedersen E, Waye KP, Ryberg JB (2010) Response to occupational noise of medium levels at four types of workplaces. In: Proceedings of INTER-NOISE, 39th international congress of noise control engineering, 13–16 June, Lisbon, Portugal
8. Golmohammadi R, Darvishi E, Faradmal J, Poorolajal J, Aliabadi M (2020) Attention and short-term memory during occupational noise exposure considering task difficult. Appl Acoust 158
9. Nassiri P, Monazzam MR, Asghari M, Zakerian SA, Dehghan SF, Folladi B, Azam K (2014) The interactive effect of industrial noise type, level and frequency characteristics on the occupational skills. Perform Enhancement Health 3(2):61–65
10. Abbasi AM, Motamedzade M, Aliabadi M, Golmohammadi R, Tapak L (2018) Study of the physiological and mental effects caused by exposure to low frequency noise in a simulated control room. Building Acoust 25(3):233–248
11. Fadda P, Meloni M, Fancello G, Pau M, Medda A, Pinna C, Del Rio A, Lecca LI, Setzu D, Laban B (2015) Multidisciplinary study of biological parameters and fatigue evolution in quay crane operators. Procedia Manuf 3:3301–3308
12. INAIL: Sistema di Sorveglianza degli Infortuni Mortali e Gravi sul lavoro, applicativo Informo web, Manuale Operativo https://www.inail.it/sol-staticinformo/informo/manuale/Manuale_Utente_InformoWeb.pdf

Mobile App Authentication Systems: Usability Issues for Cyrillic Alphabet Users (Pilot Study)

Aleksandr Volosiuk, Semyon Kuznetsov, Svetlana Podkolzina, and Evgeniy Lavrov

Abstract According to World Advertising Research Center 72% of internet users worldwide will solely use smartphones to access the web by 2025 [1]. Nevertheless, developers continue to design authentication systems within the desktop computer paradigm. The paper aims to assess the usability of authentication systems interfaces in smartphones used by Cyrillic alphabet users. The authors study authentication practices, among which password authentication is the most common. Then, the authors identify the features of creating passwords by Cyrillic alphabet users and discuss the usability issues of existing authentication tools, proposing experiment design for further study.

Keywords Usability · Authentication systems · Cyrillic passwords · Mobile ergonomics

1 Introduction

In recent years, interest in the usability of mobile authentication systems has noticeably increased. However, most of the recent studies consider authentication to the mobile device itself, and not to the mobile applications [2–5]. We also could not find any studies of mobile authentication in context of Cyrillic alphabet usage.

Among the authentication methods the most studied is a password authentication, usability of which has long been questioned [6, 7]. Despite this, a password authentication is probably the most popular authentication method to this day even

---

A. Volosiuk (✉)
St. Petersburg Electrotechnical University, St. Petersbrug, Russia
e-mail: aavolosiuk@gmail.com

A. Volosiuk · S. Kuznetsov · S. Podkolzina
ITMO University, St. Petersbrug, Russia

E. Lavrov
Sumy State University, Sumy, Ukraine

D. Sumpor et al. (eds.), *Proceedings of the 8th International Ergonomics Conference*, Advances in Intelligent Systems and Computing 1313,
https://doi.org/10.1007/978-3-030-66937-9_26

for mobile devices. Some of our previous studies discuss the relationship between ergonomic quality of systems and their overall performance [8–10].

In the current pilot study the authors aim at studying authentication practices in mobile context among Cyrillic alphabet users of different age, and laying the foundation for further usability studies of the revealed usability issues.

Thus, we propose the following three basic hypotheses:

- H.1—The most used method to authenticate in mobile devices context is password authentication;
- H.2—The passwords generated by Cyrillic alphabet users have specific features;
- H.3—Existing authentication tools do not support the features of Cyrillic passwords.

2 Methods

To study authentication practices the authors chose the survey method. For younger respondents we prepared a questionnaire, while with senior respondents we found it more suitable to use the semi-structured interview method.

The authors defined the target audience for the questionnaire as people from 16 to 35 years old. The questionnaire was conducted by using the Google Forms online service, the authors distributed the link to the questionnaire by sending messages to chats with relevant age group users (student, work and school chats). As for the senior subjects, the authors interviewed their relatives and friends.

The questionnaire consisted of the following questions:

- Age;
- In what types of applications do you need to authenticate most often?
- Which of the authentication methods do you use more often?
- Which one do you think is the most convenient?
- Which one do you think is the most inconvenient?
- Do you use a password to authenticate to the mobile device itself?
- Do you use Cyrillic words typed in English keyboard layout as passwords?

To further study the H.2, the authors analysed a leaked database consisted of 469536 email-password pairs. The database had a valid email-password pair for one of the authors' email address, which proved the database credibility.

The authors analysed the database in an automated way using a Python Programming Language in the following order:

- Domains of countries using Cyrillic alphabet (namely .ru, .ua, and .by) were selected from the database. This made possible to perform a comparative analysis of password frequencies to reveal possible features in generating password.

- Passwords have been categorized based on the character composition:
  - Numeric passwords—consisted of numbers, but not letters;
  - Letter passwords—consisted of letters, but not numbers;
  - Mixed passwords—consisted of both letters and numbers;
  - Other passwords—consisted of at least 1 special symbol.

  All the letter passwords, mixed and other passwords consisted of at least 3 letters were merged in "Possible Word" group.
- All the passwords from "Possible Word" group matching words in the British dictionary were removed from the further study.
- All the remained passwords in "Possible Word" group were transliterated (each character has been replaced with the corresponding Cyrillic keyboard character), and filtered by having at least one consecutive vowel and consonant (or vice versa) pair.
- The resulting list was compared to Russian words dictionary and analysed manually.

The authors used an expert assessment method to analyse the correspondence of authentication tools to the features of Cyrillic passwords. The authors themselves acted as experts with the required qualifications.

3 Results

We received 110 answers to the questionnaire. 99% of subjects were from 18 to 26 years old, and almost 60%—from 21 to 24 years old, mainly—students in different areas (Fig. 1). Most of the authentication cases take place in banking service applications (3) and social networks (1), less often—messengers (2), shops (6), and games (4), (Fig. 2).

The most frequently used authentication method is a password authentication (2), which is, at the same time the respondents consider the most inconvenient, while the most convenient for the responders is authentication through a third-party service (1). The other two presented options were authentication through e-mail confirmation (3) and authentication through message/call confirmation (4) (Figs. 3, 4 and 5).

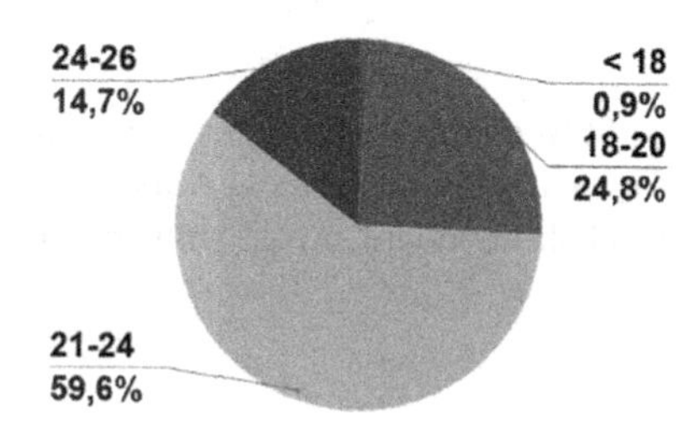

Fig. 1 Age groups of the respondents

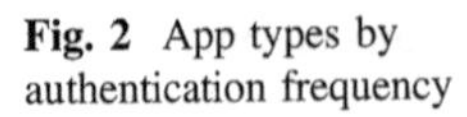
Fig. 2 App types by authentication frequency

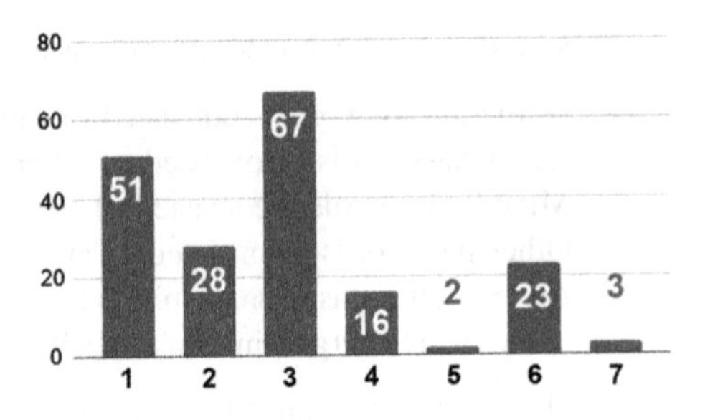

Fig. 3 Authentication methods by usage frequency

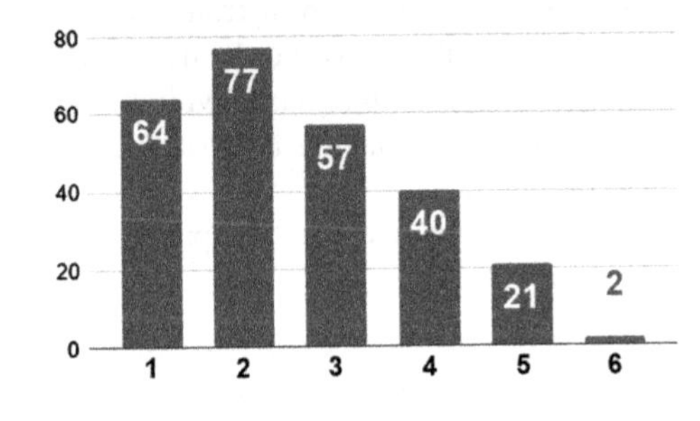

Fig. 4 Authentication methods by convenience

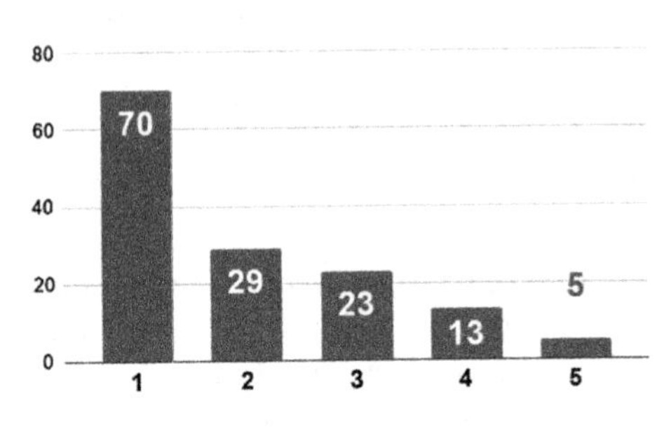

Fig. 5 Authentication methods by inconvenience

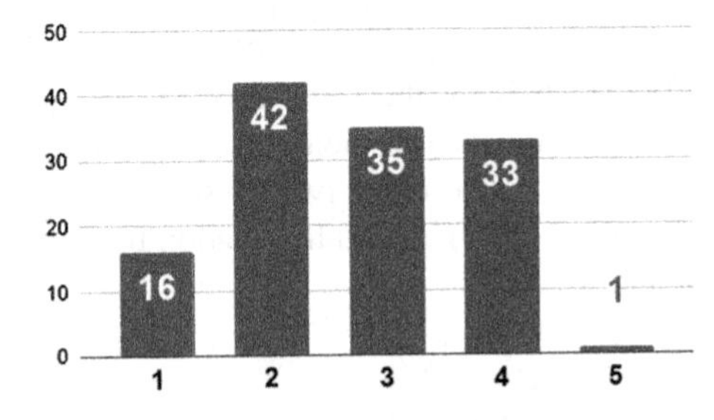

56% of the respondents use a password (not PIN) as an authentication method for their mobile devices. Almost 40% of the subjects use Cyrillic words typed in English keyboard layout (transliterated words) as passwords (Fig. 6).

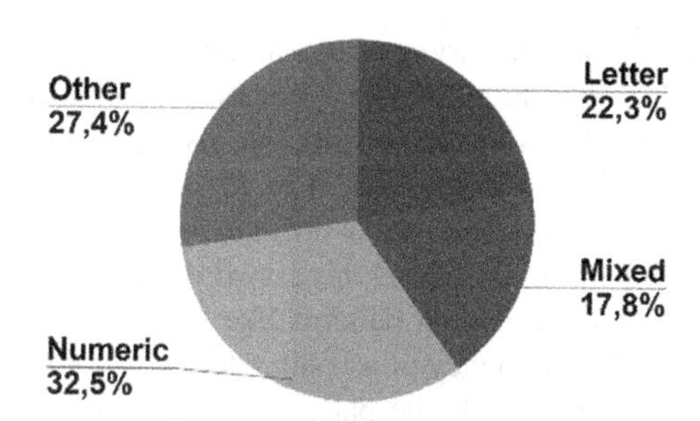

Fig. 6 Authentication methods by inconvenience

The authors interviewed 11 people from 56 to 75 years old. The interview results did not contradict the survey results and revealed the password authorization to be the most commonly used by senior people. However, for the interviewed users all the authentication methods seemed complicated.

The analysed leaked passwords database contained 469,536 records (email—password pairs). The number of records consisted of emails registered in ".ru", ". ua" and ".by" domains was 1,384. The Fig. 6 shows the distribution of passwords by character composition.

The number of "Possible word" passwords (consisted of at least 3 letters) were 835. Then, the analysts transliterated the "Possible word" passwords and kept only the words consisted of at least one consecutive Cyrillic vowel and consonant (or vice versa) pair. There were 693 such words. After manually processing the list, the analysts discovered 52 actual Cyrillic words, considered as transliterated passwords. Which accounted for 6.2% of "Possible word" passwords number.

The most common transliterated Cyrillic words used as a password were:

- "пароль" – "password";
- "привет" – "hi"
- "дмитрий" – Dmitry (common Russian name).

4 Discussion and Conclusions

Both the survey results and the interview results proved the H.1 that the most commonly used authentication method in mobile device context is still the authentication by password (Fig. 3). At the same time, the survey revealed that it is the most inconvenient method in comparison to authentication through a third-party service (like Facebook, for example), authentication through e-mail confirmation, and authentication through message/call confirmation. The interview revealed that all existing authentication methods seem uneasy for senior people, which motivates to conduct a further usability study.

Both the survey results and the leaked password database analysis confirmed the H.2 that the passwords generated by Cyrillic alphabet users have specific features, namely using the Cyrillic words typed in English keyboard layout (transliterated words) as passwords. The ratio of transliterated passwords to total number of "Possible word" passwords accounted for 6.2%, which might look relatively small, but the most common passwords from the "Possible word" list consisted of not actual words, but patterns like "qwerty", "qaz", "zxc" and others. To calculate the actual frequency of transliterated passwords of all the "Actual word" passwords one should exclude the patterns, which is a task for a further study.

To further test the H.2 authors checked the frequency of passwords "gfhjkm" (transliterated "пароль"—"password") and "ghbdtn" ("привет"—"hi") in the test database consisted of 469,536 records. The password "gfhjkm" appeared 18 times in the database, the password "ghbdtn"—5 times. Further analysis showed that in the Pwned Passwords database [11], consisted of 572,611,621 real world passwords previously exposed in data breaches, the password "gfhjkm" appeared in 267,804 leakages.

The situation with transliterated passwords is not unique for Cyrillic alphabet users. The same idea lays behind a seemingly random and strong password "ji32k7au4a83", which appears in at least 158 different leaked password databases. The point is that "ji32k7au4a83" when transliterated by Zhuyin (Chinese transliteration system used in Taiwanese Mandarin) returns "我的密碼" literally meaning "my password". At the same way transliterating the word "password" by Zhuyin returns "au4a83", which appears as a password in at least 1,548 leakages.

Having revealed the wide usage of transliterated words as passwords the authors analysed the existing tools to input passwords in mobile applications.

The figure below shows a classic keyboard used in countries using Cyrillic alphabet. The buttons of the keyboard have both Latin and Cyrillic characters (Fig. 7).

Typing transliterated words by means of a keyboard like this does not create any inconveniences since one always perceives both alternative options for each button.

Keyboards on mobile devices repeat such a keyboard in the characters arrangement but have a significant difference—only one language layout can be displayed at a time (Figs. 8 and 9). Thus, establishing a correspondence between

Fig. 7 A keyboard example with both Latin and Cyrillic letters

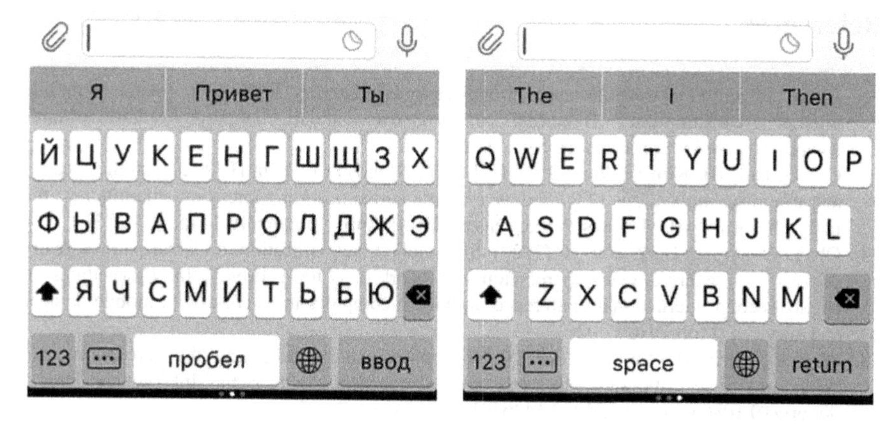

Fig. 8 iOS keyboard example in Russian and English layouts

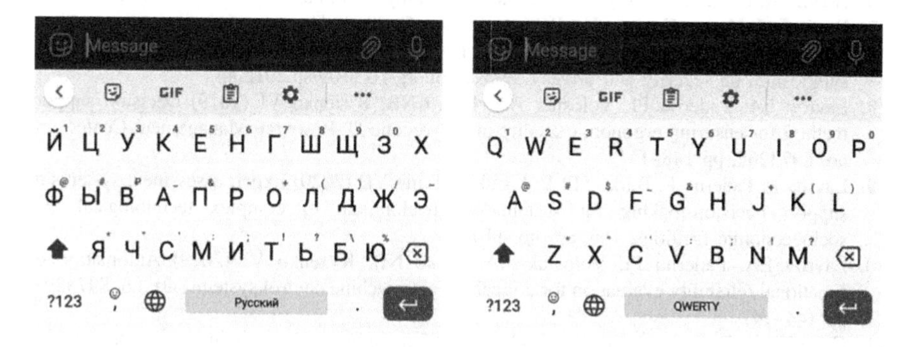

Fig. 9 Android keyboard example in Russian and English layouts

the Cyrillic and Latin symbols is much more difficult, which obviously leads to a significant increase in the time spent and number of errors when typing transliterated words. Which proves the H.3 that existing authentication tools do not support the features of Cyrillic passwords. Conducting a quantitative analysis of difficulties in typing transliterated words on mobile devices is a task for further research by the authors.

The revealed findings show an urgent need for mobile devices and applications developers to take into account the features of non-Latin script users in designing authentication solutions.

For further research the authors propose a usability study on how the Cyrillic alphabet users perform authentication process in different mobile applications.

References

1. Almost three quarters of internet users will be mobile-only by 2025. An overview of mobile internet consumption forecasts, https://www.warc.com/content/paywall/article/warc-datapoints/124845. Accessed 21 Sept 2020
2. Cho G, Huh JH, Kim S, Cho J, Park H, Lee Y, Beznosov K, Kim H (2020) On the security and usability implications of providing multiple authentication choices on smartphones: the more, the beter? ACM Trans Privacy Secur 23 (4), https://doi.org/10.1145/3410155
3. Obada-Obieh K, Beznosov H (2020) SIs implicit authentication on smartphones really popular? On android users' perception of "smart lock for android". In: Proceedings of twenty-second international conference on human-computer interaction with mobile devices and services (MobileHCI 2020)
4. Wolf F, Kuber R, Aviv AJ (2018) An empirical study examining the perceptions and behaviours of security-conscious users of mobile authentication. Behav Inf Technol. https://doi.org/10.1080/0144929X.2018.1436591
5. Kraus L, Schmidt R, Walch M, Schaub F, Möller S (2017) On the use of emojis in mobile authentication. IFIP Adv Inf Commun Technol 265–280. https://doi.org/10.1007/978-3-319-58469-0_18
6. Adams A, Sasse M (1999) Users are not the enemy. Commun ACM 42(12):41–46
7. Bonneau J, Herley C, Oorschot PC van, Stajano F (2012) The quest to replace passwords: a framework for comparative evaluation of web authentication schemes. In: 2012 IEEE symposium on security and privacy. https://doi.org/10.1109/sp.2012.44
8. Lavrov EA, Paderno PI, Volosiuk AA, Pasko NB, Kyzenko VI (2019) Decision support method for ensuring ergonomic quality in Polyergatic IT Resource Management Center, art. no. 8973265, pp 148–151
9. Lavrov E, Paderno P, Burkov E, Volosiuk A, Lung VD (2020) Expert assessment systems to support decision-making for sustainable development of complex technological and socioeconomic facilities, 166, art. no. 11002
10. Lavrov EA, Paderno PI, Volosiuk AA, Pasko NB, Kyzenko VI (2019) Automation of functional reliability evaluation for critical human-machine control systems, art. no. 8973294, pp 144–147
11. Have I Been Pwned resource to quickly assess if an online account having been compromised in a data breach, https://haveibeenpwned.com/Passwords. Accessed 21 Sept 2020

Proposal of an Ergonomic Interface for Supervision and Control of an Automated Shunting Device

Adrian Wagner, Hrvoje Haramina, and Frank Michelberger

Abstract In rail freight transport, certain processes require shunting activities to form or dissolve freight trains. In the present form, this requires shunting staff to carry out these physically demanding tasks. In spite of all precautions, accidents at work occur during such activities, which can be accompanied by serious injuries or even death. One approach to prevent such accidents will be the use of automated shunting units and an automatic uncoupler. In this paper the process of an automated system is sketched. It describes the responsibilities relating to the controlling and supervision staff of the process. It is shown how the flow of information between the parties involved can take place. Results of this research show how the proposal of an ergonomic interface for supervision and control of an automated shunting device can be sensibly integrated into the existing operation of marshalling yards. The proposal offers improvement of railway staff safety through reducing dangerous operating procedures and help to reduce the future lack of available staff.

Keywords Shunting process automation · Railway staff safety · Shunting interface · Ergonomics

1 Introduction

With regard to the plan of 30% CO2 reduction proposed by the EU Commission for the period 2019–2030, climate-friendly freight transport is a relevant component. Therefore, the railway can use its ecological system advantage in freight transport,

A. Wagner (✉) · F. Michelberger
Carl Ritter Von Ghega Institute, St. Pölten University of Applied Sciences, St. Pölten, Austria
e-mail: adrian.wagner@fhstp.ac.at

F. Michelberger
e-mail: frank.michelberger@fhstp.ac.at

A. Wagner · H. Haramina
Faculty of Transport and Traffic Sciences, University of Zagreb, Zagreb, Croatia
e-mail: hharamina@fpz.unizg.hr

D. Sumpor et al. (eds.), *Proceedings of the 8th International Ergonomics Conference*, Advances in Intelligent Systems and Computing 1313,
https://doi.org/10.1007/978-3-030-66937-9_27

but the quality of a transport chain is only as good as each individual link within the overall process, so all process components must be analyzed to determine whether there is potential for optimization. Optimizations, in the form of time or cost savings, represent an incentive for freight forwarders when choosing the means of transport [1].

For that marshalling yards are a starting point for optimization; these stations represent a significant hub for freight wagons in single wagonload traffic along the transport chain. The technologies used at these stations are proven, but require a high level of personnel commitment. Thus, many actions, such as unhinging the screw coupling of rolling freight wagons, are carried out by employees through physical activity at any time of the day or year. In spite of all precautions, accidents occur during such work, which are accompanied by serious injuries or even death [2]. Investigations into the use of new technologies have resulted in approaches to adaptations at marshalling yards. In this paper, two of these approaches, automated shunting and automated uncoupling and their effects are considered. In this respect, it should be considered which process-related changes could result and which responsibilities arise in the control and monitoring. In addition, the necessary information is to be considered and how the ergonomic design of an interface is possible. For the research Austrian operational procedures are used and analyzed as a basis for the elaboration.

1.1 Current Situation

Freight wagons or groups of wagons are bundled to exploit synergies in single wagonload traffic. In this process, goods from different regions which cover the same transport route between two hub stations are combined at hub stations. This sorting takes place in shunting nodes or marshalling yards. Marshalling yards are characterized by an entry group (receiving yard), a directional track group (classification bowl) and an exit track group (departure yard). Incoming trains are prepared in the receiving group and the individual freight wagons or wagon groups are rolled over a hump into a specific track in the set of sorting sidings. These tracks have a point code and thus a destination station can be assigned to these individual tracks [3]. Freight wagons or groups of wagons with the same destination station, or intermediate destination, are thus lined up together and prepared for departure in the departure yard. Figure 1 shows an overview of these groups.

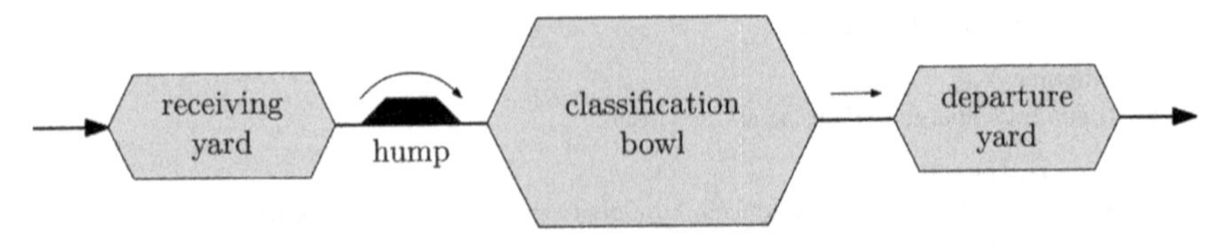

Fig. 1 Typical layout of a marshalling yard. *Source* Taken from Marton et al. [4]

Marshalling yards differ worldwide, but also within the individual countries, in part due to the most diverse infrastructure, types of operation, but also due to the technologies used. The basic processes and the topology are comparable.

1.2 Processual Procedure

In this paper the focus of the measures lays on the receiving yard of marshalling yards. This is due to the fact that in the research projects carried out the investigations take place in the mentioned area. The consideration of the allocation group and the departure sidings are considered in other papers. However, the necessary functionalities can already be shown in this partial consideration. The handling process of a freight train starts already before its arrival at the station. Here the focus is on the planning with the dispatching system. In addition to the dispatching system, an operation system is used, the interface to external systems is an interface server. In Austria, the system primarily receives the necessary information via the RZÜ (computer-aided train monitoring) and the INFRA-TIS (Infrastructure Transport Information System). This results in the dismantling list with the separation points and the outgoing train list [5]. Initially, after the arrival of the incoming train, the wagons are secured against unintentional unrolling; the choice of the securing means has to be made according to the locally valid regulations. This safety device are brake shoes which are applied by a switchman, in addition, incoming trains are also braked by full braking. Afterwards the locomotive is pulled off and the separation points are prepared. During this process the couplings are lengthened. From this point on, the wagon group can be coupled by means of a shunting vehicle with a shunting coupling and the securing devices can be removed. Afterwards, the train is shifted towards the hump. In front of the hump, on the approach tracks, switchmen use a bar, a long metal rod, to release the screw couplings while the train is passing. The wagons then roll through the gravity into the allocation track [3]. Figure 2 shows an overview about the different processes on the shunting yard.

1.3 Deployment of Staff

The division of the employees, who are necessary for the accomplishment of the operational business, can be assigned in three groups: The driving personnel, the signal cabin personnel, personnel in the track area. In addition, further occupational groups are required to maintain operations in the long term, primarily maintenance personnel in various disciplines. In this paper the main focus of the work is on the shunting personnel, who are responsible in the receiving yard for securing the wagons, making the screw couplings long and unhooking them. This is a physically demanding and not harmless job [6]. The average age of employees of the Austrian

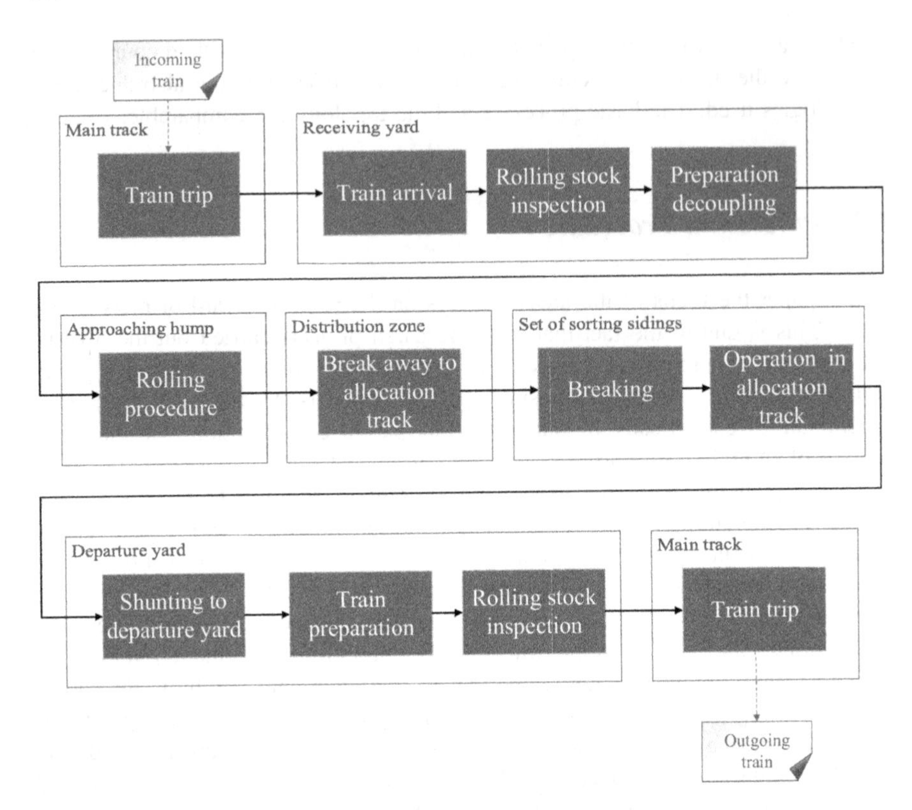

Fig. 2 Process overview of a marshalling yard. *Source* Modeled with MS Visio

Federal Railways is 46 years, whereby the average age of retirement has increased by 1 year between 2013 and 2016 and is now between 60.2 and 61.1 years, depending on the employment contract. Retirement due to illness will take place earlier accordingly [7]. A psychological study of train drivers in Austria analyzed different aspects in three age groups (20–25, 26–35 and 36–59 years). Among these aspects were the median reaction time and the reaction to stimuli. It was found that older persons react significantly later to stimuli and that the reaction time is also longer [8]. A similar picture can be seen in the accident statistics of the AUVA (General accident insurance institution), which manages the records of occupational accidents in Austria. In summary, most accidents at work occur in freight transport, and the majority of incidents involving locomotion (including with means of transport—walking, driving, riding) in the period from 2010 to 2019 occur in the age group from 45 to 54 years [9]. In addition, the search for new employees as a switchman is a challenge for human resources managers, as fewer and fewer people want to take on this job. Therefore, these aspects are given special emphasis in the research projects.

2 Methods

First of all, an open-ended survey of existing technologies is carried out in an overview section and the corresponding processes are documented. In addition, a mix of methods is applied, whereby on the one hand, a systematic literature search and the research of relevant operational framework conditions are carried out. On the other hand, expert interviews are conducted in order to conduct a complete examination. To complete the questionnaire the method of scientific observation is applied. The surveyed processes are mapped using the BPMN standard. To examine the personnel structure, existing statistics are used and interpreted. The knowledge gained from previous research projects, such as the decoupling robot, is implemented in the existing processes and by comparing them with the existing processes, the deviations are mapped. From this data, the processes are mapped again, whereby requirements can be created for the interface by means of requirement engineering.

3 Results

In an automated scenario of the entry group of a marshalling yard, the possibility of different approaches to adaptation is shown. If one considers at present possible application of the independent rolling operation by given driving commands from the operation system, here the component of the bypassing of freight cars and also the uncoupling of car groups can be automated. The decoupling robot under consideration is a research object that was developed by an Austrian research consortium in projects funded by the FFG. The automated shifting unit is an object that is currently being considered in various research projects. Therefore, in this chapter the possible process sequence is discussed and presented in relation to these two changes. The comparison is drawn with the initial situation described in the introduction chapter.

3.1 Sequence of an Automated Process

In the scenario presented, first of all the planning preparation for the access remains unchanged. The incoming train is assigned a route by the dispatcher to the entry track group. There, the driver stops the train according to the valid description of the station. An employee secures the train against unrolling by means of restraining shoes. The traction unit is uncoupled by the uncoupling robot and can be removed. By means of an interface the employee reports the wagon train to the system ready for the preparation of the separation points. The couplings are lengthened at the corresponding separation points. At the same time a automated locomotive moves

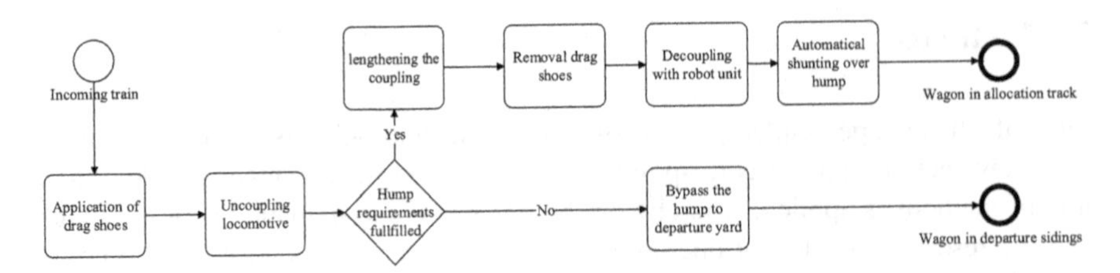

Fig. 3 Simplified overview of the rolling process in the receiving area of a shunting yard. *Source* Modeled with MS Visio

to the rear of the freight cars. This couples automatically by means of a shunting coupler. The employee now removes the safety devices and bleeds the brake line. He then withdraws to a place where employees can stay (e.g. shunting platform) or moves to the next wagon group to be unrolled. In the meantime, he reports that the first wagon train is ready to be unrolled. The shunting locomotive receives a departure order from the operating system with a target speed. At this speed, the wagons pass the decoupling robot, which uncouples the couplings as they pass by. The wagons that have passed the highest point of the decoupling hill roll automatically into the specified directional tracks. Before the last wagon train rolls off, the shunting coupling is released from the shunting coupler and these wagons also roll towards the directional track. The shunting locomotive now bypasses further wagon sets and the process is repeated. A protection of the wagons against de-rolling without the participation of employees was not considered in this application. In summary, the securing of the stationary freight wagons and the confirmation in the system of this condition must be done manually. The main advantage is therefore that manual activities by the staff are massively reduced. Especially activities where employees are located between the buffers of the freight cars could be completely avoided in this scenario.

Figure 3 serves as an overview and provides a highly simplified excerpt from the created process sequences.

3.2 *Responsibilities*

There are two possible cases of responsibility:

- Normal case:
  - Control by the traffic control center
  - Automated
  - Switchman on site for placement of drag shoe.

- Malfunction:
  - In case of error, failure of the system or partial components
  - For unusual operating actions
  - Partial or complete operating actions on site.

As a normal case, the responsibility and monitoring of the system should be carried out by the traffic control center. Only the responsibility for placing and removing the drag shoes should be borne by the employee on site.

In case of an error, information about the operating status should continue to be sent to the traffic control center. However, the traffic control center should not be able to carry out any further operations, as personnel may be located in the track area. Thus, comparable to an electronic shunting road request, a certain area of the entry group should be transferred to the responsibility of the personnel on site. Now the personnel on site can carry out all operating actions or perform them manually and also are responsible for the whole procedure. As soon as these activities have been carried out on site, the employees return the authorization, then the traffic control center can carry out operating actions again.

3.3 Information Transmission

The transmission of the information must comply with the current standards used in the field of traffic control and postponement. Therefore, a Fail—Safe principle is to be considered. It is to be controlled therefore in regular time intervals independently by the system whether a connection between all components exists. If it comes to the failure is in each case a roll stop or a general emergency stop of the journeys to be accomplished. In addition, common standards such as EIRENE (European integrated railway radio enhanced network) could be considered for the transmission of information in order to define corresponding requirements.

3.4 Interface Proposal

If an interface is used to operate automated shuttle units, to take over control in case of control or failure or to set manual actions by shuttle personnel, it must meet various requirements. Analogous to portable remote-control units for radio remote controlled traction units, the interface must be carried by the operator at all times. Translated to the described application case: If the interface is monitored or controlled by the control center, the operator is at his workplace when operating the interface. The same applies analogously to employees on site with mobile operating devices. Therefore, this device must meet the requirements of occupational safety and the operability must be adapted to the conditions on site. In order to avoid errors caused by operator actions, the interface should have a user interface for which all possible operator actions are always visible and no sub-selection is required. In addition, the status of the equipment should be immediately recognizable and already interpreted by the system to avoid possible interpretation errors. All operating actions should be easy to perform, but must comply with the requirements of the railroad industry, therefore confirmation actions have to ensure that an entry was not made unintentionally [10]. So, the common usability requirements have to be met, the design has to be tailored to the operator.

4 Conclusion

The automation of physically hazardous work brings advantages for the people involved, but changes in personnel deployment lead to altered conditions. Considering the fact that the recruitment of personnel for shifting activities is causing more and more difficulties for those responsible, a change in the job description can change this situation. By minimizing or eliminating physically demanding work, new stimuli for recruiting young employees can be created. Equally, however, such changes should involve the relevant professional associations and trade unions at an early stage in order to directly involve the occupational group that is directly affected by changes. This way, employees can participate in the development of use cases at an early stage and in the further course of usability tests and contribute their experiences from the daily work environment. In this paper the consideration is limited to one region of the marshalling yard. The results will be further discussed in an overall view by means of further research. In particular, models and simulations will be developed to complete the view. The question of how to deal with the positioning and removal of protection means against de-rolling could not be answered in the present study. In addition, LCC calculations for the investments and a technology assessment are to be carried out. With this systemic interpretation, it is possible to make assessments of the overall system and its effects.

References

1. Henrici T (2018) EU mobility package: CO2 limit for trucks, support for autonomous driving (EU-Mobilitätspaket: CO2-Grenze für LKW, Förderung für autonomes Fahren). RailBUSINESS 21/18, p 2
2. Zeitung K (Hg.). Lethal accident in WIEN-KLEDERING (Tödlicher Unfall in WIEN-KLEDERING). URL https://www.kleinezeitung.at/oesterreich/4679720/Toedlicher-Unfall-in-WienKledering_OeBBArbeiter-von-Lok-ueberrollt. Last accesed 2020/05/06
3. Michelberger, Fabian, Jungwirth, Kalteis, Merschak, Stadlmann, Stock, …, Wagner (2018) Study for an innovative, low-noise and low-wear braking system on the hump yard—StilvA (Studie für eine innovative, lärm- und verschleißarme Bremsung am Abrollberg StilvA), p 20–25

4. Marton P, Maue J, Nunkesser M (2009) An improved train classification procedure for the hump yard lausanne triage. In: 9th workshop on algorithmic approaches for transportation modeling
5. Sturm P, Krims V. Modernization of marshalling yard Graz (Modernisierung Verschiebebahnhof Graz). SIGNAL + DRAHT (105) 3/2013, p 24–26
6. Zellner C, Stadlmann B, Egger M, Hattinger M (2017) Automated decoupling of screw couplings in shunting (Automatisiertes Trennen von Schraubenkupplungen im Verschub. Trennvorrichtung), p 2
7. ÖBB (ed). Fact is: the retirement age of ÖBB employees has been rising continuously for years (Fakt ist: Das Pensionsalter der ÖBB-Mitarbeiterinnen und Mitarbeiter steigt seit Jahren fortlaufend an). https://konzern.oebb.at/de/ueber-den-konzern/fakten/pensionen. Last accessed 2020/09/25
8. Kanzler B, Nechtelberger A (2012) The psychological aptitude of train drivers and operations staff in Austria (Die psychologische Eignung von Triebfahrzeugführern und Betriebsbediensteten in Österreich). Österreichische Zeitschrift für Verkehrswissenschaft ÖZV. 59. Jahrgang, 3. Heft, p 9–12
9. AUVA (ed) (2020) Work accidents (excluding commuting accidents) recognized by VAEB 2010–2019. (Anerkannte Arbeitsunfälle (ohne Wegunfälle) der VAEB 2010–2019)
10. Preim B (1999) Development of interactive systems. Basics, case studies and innovative fields of application (Entwicklung interaktiver Systeme. Grundlagen, Fallbeispiele und innovative Anwendungsfelder). Springer, Berlin Heidelberg, p 24

Printed in the United States
by Baker & Taylor Publisher Services